ANIMAL SEX CONTROL

A Practical Guide for Breeders

By

CARL WARREN, B.A., B.S., M.S.J.

Edited By

THOMAS DURFEE

Director of the Applied Research Laboratories, Inc., Dayton, N. J.

ILLUSTRATED

1941

A FAMOUS SEX-CONTROLLED ANIMAL

This rat, No. A2416, was one of the first mammals to prove the acid-alkali method successful in America. Treated with lactic acid, she bore a litter predominantly female. Then, treated with soda, she bore a preponderance of males.

PREFACE

The new science of sex control as presented in this volume already is demonstrating its tremendous economic value to the breeders of animals raised for profit. It will rank in the future as a contribution of the first magnitude to human happiness.

While the method was discovered in Europe the fullest credit for lifting the truth from obscurity, perhaps saving it from oblivion for centuries, must be awarded to a great newspaper—*The New York Daily News.* This book developed naturally and directly from the enterprise of *The News* which planned, financed and drew world-wide attention to the first successful experiments in America.

The author wishes to acknowledge specifically the permission given to him by J. M. Patterson, president of *The News,* to use its copyrighted data, charts and pictures appearing in every chapter. The cooperation of Richard Clarke, managing editor, and Ruth Reynolds, special writer, also is gratefully acknowledged.

Dr. Leon J. Cole, nationally known geneticist of the University of Wisconsin, gave valued suggestions but the statements and conclusions reached in the book are not necessarily those of Doctor Cole. The author and the editor, Thomas Durfee, also appreciate the assistance of Nathan Kaplan, biologist, and M. Wolfe Lerner, bacteriologist, of the Applied Research Laboratories, Inc.

C. W.

CONTENTS

ILLUSTRATIONS

ANIMAL SEX CONTROL

Animal Sex Control

I

SUPERSTITION AND SCIENCE

The Basic Facts—Sex control or predetermination—the science of producing male or female offspring at will by artificial means—today is an actuality. Within a short time it will be commonplace.

The method in its simplest form consists of a painless treatment with one of two cheap and harmless substances—bicarbonate of soda and lactic acid. It has worked in a majority of cases with rats, rabbits, cattle, horses and other mammals, including man. The evidence of its success is overwhelming and conclusive.

Hundreds of breeders already are using the method and reaping the economic benefits. A multitude of others are eagerly demanding more information and there is every reason to believe that the next few years will see its perfection for human use and adaptation to every mammal raised for profit.

Modern sex control came forth only after centuries of guesswork and slow groping through the darkness of ignorance and superstition. It still is in its infancy, for it was only 20 years ago—a comparative instant in history—that a German biologist accidentally found the key to the problem.

Nevertheless, tremendous progress has been made. Full confirmatory experiments recently have been completed in America under sponsorship of *The New York Daily News* at the Applied Research Laboratories, Dayton, N. J., and the new science now is in the practical education stage. Like other scientific achievements, such as aviation, it faces typical obstacles in the early stages of development. Because of its revolutionary significance it appears incredible to many persons and because it artificially upsets what they regard as laws of nature, others view it with forebodings. As a matter of fact, it is no more miraculous than flying and it promises to be not only a boon to breeders of domesticated animals but, when fully

Figure 1.—SUCCESS CROWNED HIS EXPERIMENTS

Thomas Durfee, director of the Applied Research Laboratories, Dayton, N. J., who supervised for The New York Daily News the first conclusive sex control tests in the United States.

accepted by human beings, a long-sought means of bringing greater happiness to mankind.

Boys or Girls to Order—As far back as written records go, and doubtless farther, men and women have fervently wished for and earnestly sought a way to predetermine the sex of their children. This desire grew out of both the practical realities and the sentimental predilections of life.

Boys were at a premium in primitive society because a male was best equipped to provide food and protection for the family group. Sons became successors to power and heirs to possessions. Similarly, modern fathers and mothers often yearn for a boy to carry on the family name and business. For reasons less practical, but nevertheless impelling, a majority of parents want girl babies as objects of affection.

Generally speaking, parents both primitive and modern desire at least one male child—preferably the first-born—and thereafter a daughter or children of both sexes. The almost universal demand is not for children of one sex exclusively, but rather for the ability to choose the sex of each child before its arrival. This demand can now be fulfilled in a majority of cases by a safe and inexpensive method of sex control.

More Profits for Breeders—The intrinsic value of obtaining the desired sex in breeding animals has been obvious to men since the first wolves were captured and trained as dogs. Heretofore, the proportion of male and female young produced by a mating has been a matter of chance. With scientific breeding commonplace, the possibilities of predetermination or control are almost boundless. Cattle raising furnishes a clear-cut example. Dairymen want a preponderance, probably 90 per cent, of heifer calves for milk. Beef raisers, on the other hand, want a preponderance of bull calves for meat. Sex control means that they may breed with more profit to themselves and at the same time produce cheaper milk and meat for the consumer.

Goat breeders, like dairymen, want females for milk. Rabbit breeders get higher prices for females used in laboratory tests. Mink and fox ranchers in some cases desire more females for breeding, while others want males because of the better quality of their pelts. Dog and cat fanciers

FIGURE 2.—SMALL ANIMALS SUBJECT TO CONTROL

Typical mammals bred for profit to which acid or alkali treatments may be applied to predetermine sex.

generally prefer males as pets. Horsemen often desire males to perpetuate a strain.

It seems likely, therefore, that the sex of calves, kids, pups, colts and kittens in vast numbers soon may be decided in advance of conception as a matter of business acumen and common sense. Beyond the immediate value of controlling the sex of an animal or litter lies the possibility of influencing other characters so important in any kind of selective breeding.

Sex Control in Folklore—The tribal witch doctor or magician with a cure for every ill and a prophecy for every event never was at a loss for taboos and concoctions helpful in arranging the sex of infants. Folklore is full of such formulas. Impelled by an innate longing for children of one sex or the other men have relied upon these false and frequently ridiculous superstitions since the dawn of history.

The ancient Egyptians believed that if a fish bone were placed on a woman's head the sex of her child would be that of the next person she met. In Tyrol, if the husband wanted a son he was advised to wear his boots during coitus. Another magic expedient to insure a boy was to smear himself with hare's blood. Biting a wife's ears during the sexual act was regarded in Modena as a sure method of getting the desired son, while in one of the Balkan states a sty on the upper eyelid was a sign that a son would be born.

In the 17th century a French physician, Felincourt, compiled a list of 262 existing theories of sex control. Many a Frenchwoman favored an egg method. She would take an egg from under a sitting hen and keep it warm in her hands until it hàtched. The sex of the chicken, she thought, would be the sex of her child.

To this day Korean women revolve a giant stone to fix the sex of their offspring. In the Pennsylvania hill-country one hears about an old, old method. The father hangs his pants on the right side of the bed to beget a son and on the left side for a daughter. It is believed in parts of Italy that the wearing of a hat at the time of intercourse is helpful in obtaining a son, while in Poland the wearing of Sunday clothes on this occasion is held to have a powerful influence to the same end.

The Logic of the Ancients—The creators of these fantastic and futile theories were by no

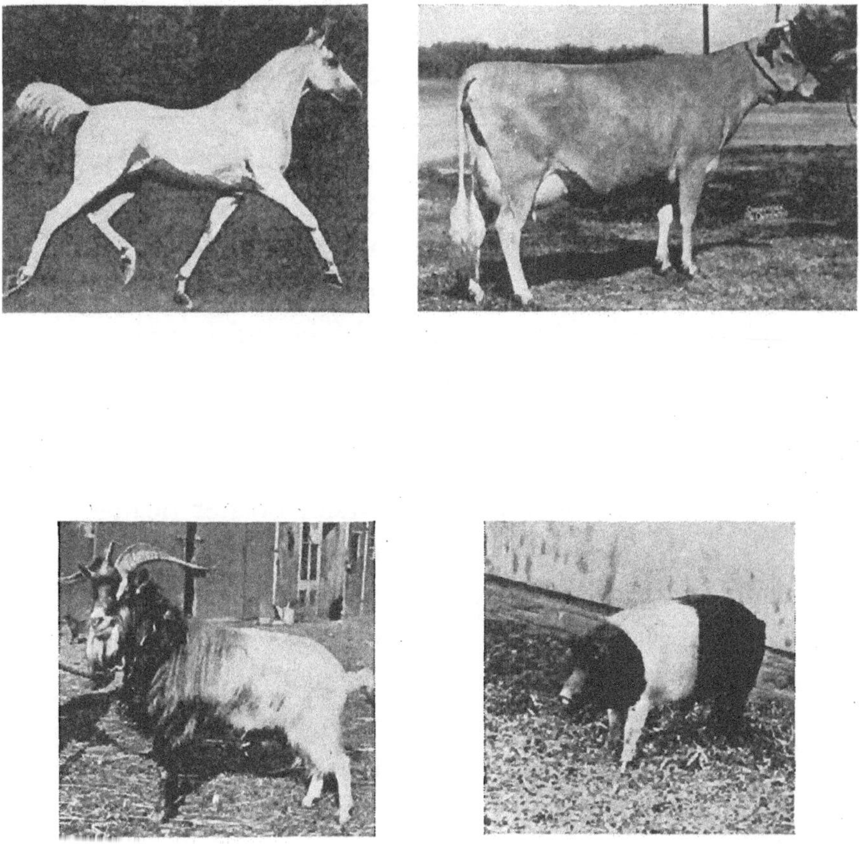

FIGURE 3.—LARGE ANIMALS SUBJECT TO CONTROL

Horses, cattle, goats and swine are among the farm animals to which sex control is applicable.

means all quacks and fools. The most intelligent men of their eras pondered and debated the problem but usually with more credit to their imaginations than to their knowledge.

To the great anatomist, Galen, founder of experimental physiology, is ascribed the origin of almost unbelievable notions about sex control. A woman could bear a male child, he suggested, if before meeting her husband she bound her right foot with the white hair ribbon of a child and a daughter if she tied a black ribbon to her left foot. Taking a cue from the Egyptian school, Galen also proposed placing a sprig of parsley on her head—the sex of the child to be that of the next person she met. Still another Galen prescription was to cover the body with goose grease and resin for three days before coitus to obtain a son.

The philosophers and doctors of Greece, relying largely upon logic and observation, connected the determination of sex more closely to bodily functions than many of the ancient doctors.

About 500 B.C. Anaxagoras started a school of thought which held that the sex of a child was determined by the testicle from which the semen flowed. Hippocrates, father of modern medicine,

FIGURE 4.—SCENE OF SEX CONTROL EXPERIMENTS

Main building of the Applied Research Laboratories at Dayton, N. J.

thought there was male and female semen and if the male semen was in greater quantity or stronger in quality a son would be born. Aristotle suggested that a north wind blowing at the time of conception would bring a boy, while a south wind would bring a girl.

These speculations were probably the shrewdest guesses about sex control prior to the modern era. Strangely enough, in the fantastic wind theory appears the first faint hint of the truth for, as will be seen later, the seasons may possibly be a slight factor in natural determination.

Modern Fallacies—While future scientists probably will find a reliable test for mammalian mothers-to-be which will reveal the sex of young before birth, they are much less likely to change the sex of young after conception has occurred. It is true that in some of the lower forms of life sex can be changed or profoundly altered after conception by external means, but in man and the higher animals such basic changes are not now possible.

This disposes immediately of a superstition oddly accepted as gospel truth by many people of our own day who believe that suggestion in some way has a positive effect on the sex of the unborn child. They think that if a prospective mother keeps concentrating on the wished-for sex, her child will be of that sex when born. Many followers of Coue, who popularized the saying, "Every day in every way I am getting better and better," held auto-suggestion to be a valuable asset to a mother hoping for either a boy or a girl.

Such a theory is sheer nonsense and in the same class with the thoroughly discredited fallacy that pre-natal impressions result in birthmarks. No mental process can change the sex of an unborn child any more than it can change the sex of a man or woman sixty years old.

The known fact that sex is determined as the sperm of the male and egg of the female unite also blasts away a belief that the mother's mode of life and diet during pregnancy—particularly that overfeeding tends to result in a girl and underfeeding in a boy—influences the sex of babies. Such a belief is contrary to unassailable scientific proof that sex is established at the moment the egg is fertilized.

Prediction and Predetermination—It should

be made perfectly clear that control and predetermination mean artificial directing or adjusting of sex by choice before conception. Natural determination refers to the way nature produces the sexes, roughly on a 50-50 basis. Prediction means announcing, with or without evidence, the sex of a child after conception but before birth.

Let us repeat the established scientific fact that the question of maleness or femaleness in mammals is resolved at the time of conception or union of the reproductive cells, not before conception and not after conception. Failure to understand this point has caused confusion from the earliest times and still gives rise to many false presumptions.

Hippocrates, for example, held that the semen determined sex, a first-rate guess for his era. This was his theory to explain natural determination. His method of prediction, wholly different, was absurd. He told the Grecian housewife to form a paste of mixed flour and milk and to bake it over a slow fire. If the paste was consumed by the blaze, he said the child already in her womb was a boy; but if it cracked and split, it was a girl.

There are almost as many superstitious ways of

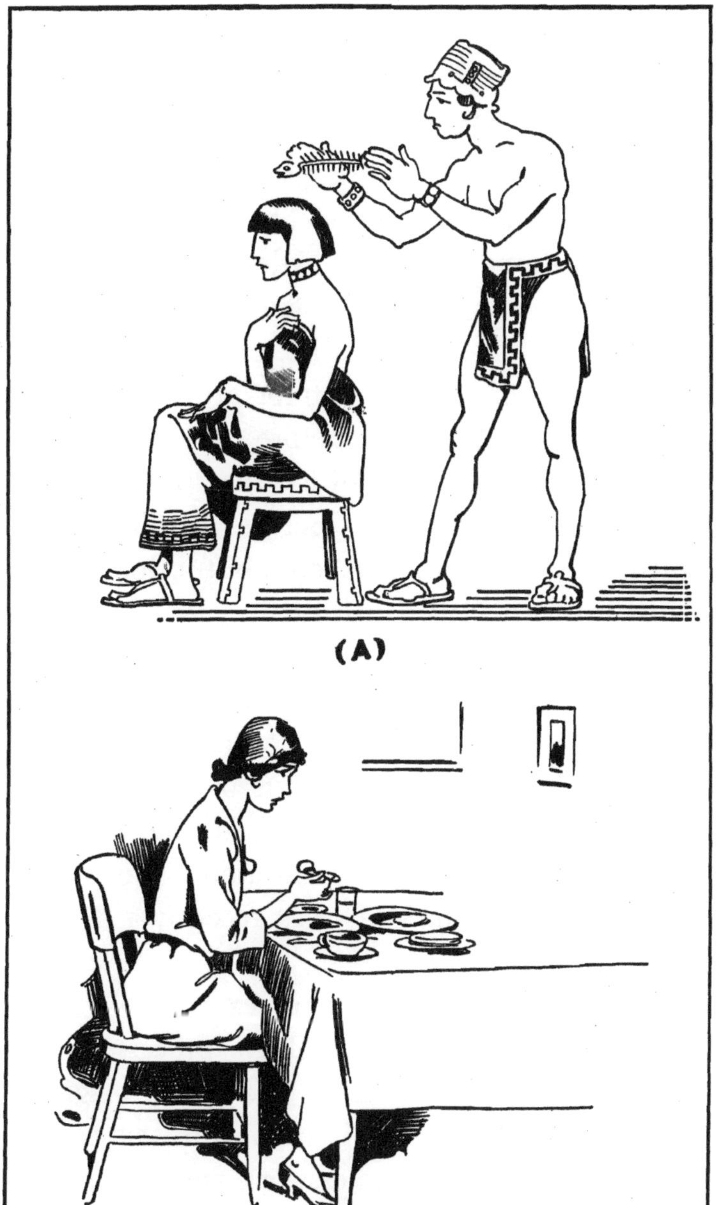

FIGURE 5.—ANCIENT AND MODERN SUPERSTITIONS

(A) The Egyptians thought a fish bone placed on a woman's head influenced the sex of her child. (B) Credulous expectant mothers today believe that proper eating habits will give them a boy or girl as desired.

predicting as of predetermining sex. None of the prediction methods has any scientific basis except insofar as prediction follows predetermination.

The person who artifically fixes sex prior to conception has reason to predict, of course, what the result will be. But he cannot prove it until after the birth. As stated before, it is entirely within the realm of possibility that a reliable method of distinguishing sex between conception and birth may be found, and soon, but it does not exist now.

The Acid-Alkali Method—Sex control from the earliest times, as we have seen, has been a fertile source of superstitions. Man constantly strives to widen the narrow limits of his knowledge. He has thoroughly explored the earth, measured the stars, analyzed the atom. Yet the generation of life remained until recently a deep mystery.

Deeper still was the secret of sex control, for knowledge of the essential biologic processes in procreation had to come first. While the origin of living matter itself still is unknown, sex control was graduated from guesswork and superstition to the status of a modern science only after the procreation processes were charted sufficiently well to know how, if not why, living matter behaves as it does.

Like many other solutions to long-buried mysteries, the soda-lactic acid sex control method proved to be basically simple. Soda, the common name for bicarbonate of soda, an alkali, is the ordinary type found on every housewife's shelf. Lactic acid, a substance not so frequently seen around the house, is nonetheless common.

However, do not jump to the conclusion that all one need do is dash off to the kitchen or medicine shelf and pour out a teaspoonful of either and then be prepared to obtain girls or boys, heifers or bulls, mares or colts. The acid-alkali method is a little more complicated than that and it cannot be fully understood without a rather detailed picture of the reproduction mechanism.

II

HOW NATURE DETERMINES SEX

The Transmission of Life—The way in which body cells grow and perpetuate the functions peculiar to organic or living beings, as distinguished from dead or inorganic matter, may be described as a rudiment of modern biology.

The cell, a minute mass of protoplasm containing a nucleus and inclosed in a covering or cell wall, is the structural unit of all living creatures. This tiny particle multiplies itself by dividing and forming new cells. Thus, life is carried on continuously. Where the first cells came from we can only speculate, but we do know how they reproduce the complex organisms we call plants and animals.

One of the simplest animals, the single-celled amoeba, reproduces itself by merely dividing into two parts, each of which becomes a full-grown amoeba. But in human beings and other forms of life, reproduction is accomplished in two steps:

(1) a dividing process by body cells to form special types of reproduction cells called gametes; and (2) the union of two gametes to form a zygote or new being.

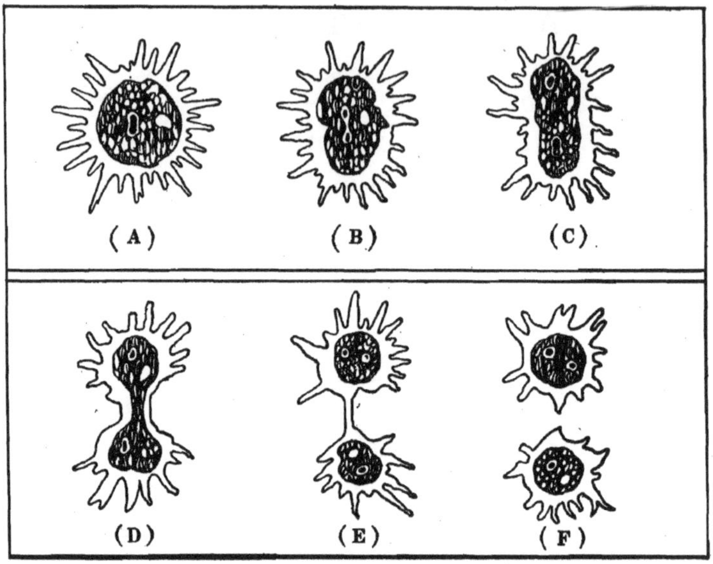

FIGURE 6.—DIVISION OF AN AMOEBA

How a single-celled animal multiples by dividing into two parts each of which grows into a new individual.

The gametes are of two kinds—the sperm (singular sperm or spermatozoon, plural sperms or spermatozoa) which is produced by and distinguishes the male and the ovum or egg cell which is produced by and distinguishes the female. While the same individual in the simpler organisms often produces both sperms and eggs, an individual in the higher groups produces only one kind of gamete and thereby its sex is designated.

The zygote or product of the merger of a sperm and an egg grows into an embryonic child, pup, calf or colt and ultimately into an adult of the same species as the parents. The zygote contains within itself potential gametes or germs of new beings.

The Sperm and the Egg—Let us now examine more closely the two reproductive cells or gametes, for upon their remarkable make-up and uniform activity rest the explanation of acid-alkali sex control.

The protoplasm of each sperm and egg contains a nucleus surrounded by cytoplasm. The nucleus, a clear grey globule which is the governing or control center of cell activity, is fed and protected by the viscous and translucent cytoplasm which envelops it.

While alike in containing nuclei and cytoplasm, the sperm and egg differ widely in other ways. The unlike contents of the nuclei, to be analyzed later, explain why one is male-determining and another female-determining. In outward appearance, too, as well as locomotion, the gametes are wholly dissimilar.

The human egg, no larger than a pin point, measuring 1/120 to 1/150 of an inch in diameter, is a thousand times bigger than the average tissue cell and at least 2,000,000 times bigger than the sperm. The egg is round or oval in shape. It often is loaded with food material, as in birds, for the development of a new individual, but in mammals, including man, the food material is less abundant since the egg so early becomes imbedded in the uterine wall where is obtains nourishment. It possesses no motile organ and cannot move by itself.

Now consider the sperm or male gamete. Invisible to the naked eye, its existence was unsuspected until a study with high-powered lenses was undertaken. In shape it resembles a tadpole with a blunt head 1/6,000 of an inch long, a middle portion or body and a whip-like filament or tail, 1/400 to 1/500 of an inch long, which constantly vibrates like a whiplash and so carries the sperm forward spontaneously. If a fresh drop of semen

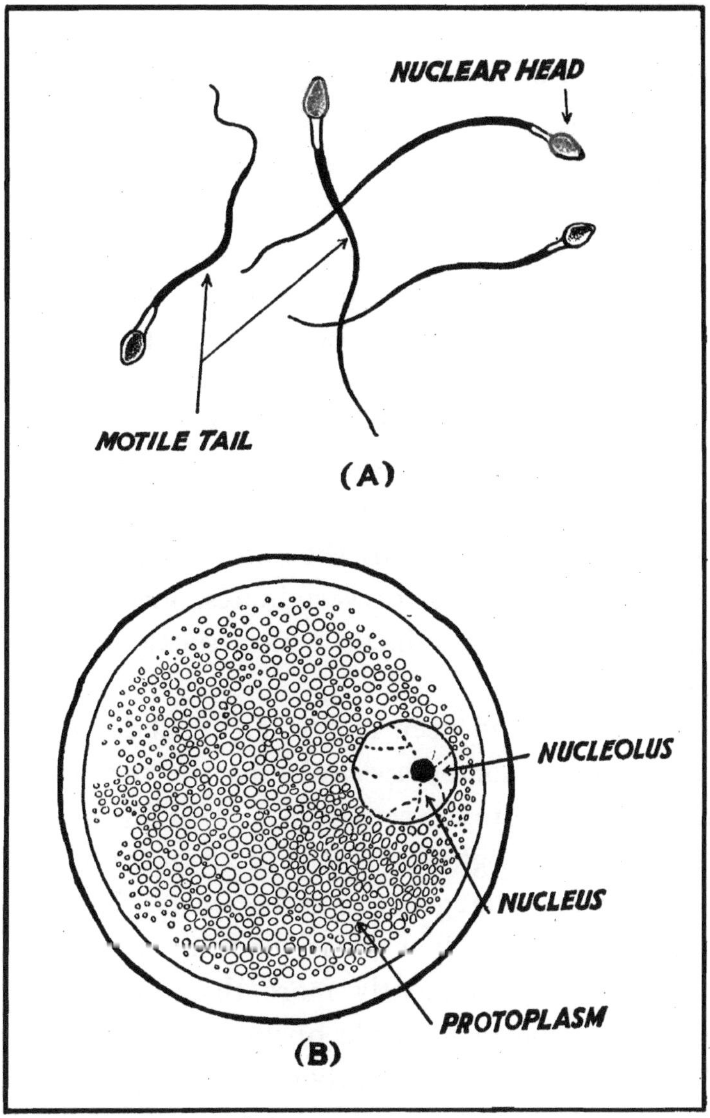

FIGURE 7.—MALE AND FEMALE GAMETES

Sketches of highly magnified sperms (A) and ovum (B) of humans.

is examined under the microscope, sperms can be seen swimming about in all directions.

When a sperm comes into contact with the surface of an egg it bores its way inward. Upon entering, the sperm nucleus swells, moves toward the nucleus of the egg and fuses with it. The egg then is fertilized and the growth of the zygote into a new individual begins.

Origin of the Egg—In order to present a clear picture of the gamete fusion or merger, the reproductive processes of human beings must be explained. They are roughly the same in all mammals, the highest class of vertebrates, including man and all other animals which suckle their young.

Out of sight and deep in the female abdomen is a pair of almond-shaped glands, the ovaries, which manufacture the eggs or ova. Usually once each 28 days a single egg ripens in an ovary of a woman between the ages of 14 and 45 years. There may be slight individual variations in the period, however, and infrequently more than one egg ripens, resulting in multiple births.

During ovulation this newly-ripened egg is discharged and, propelled in fluid currents by waving hair-like cilia, makes its way down a passageway or oviduct—one of the two Fallopian tubes. These ducts, about 4 inches long and the diameter

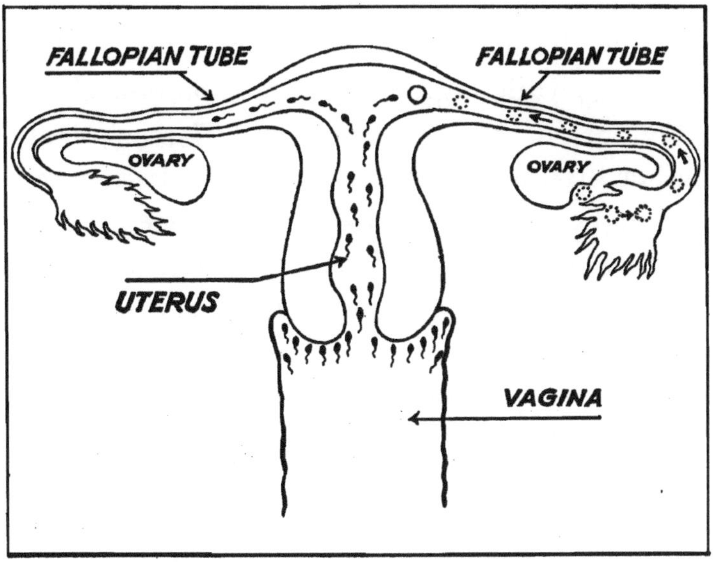

FIGURE 8.—FEMALE REPRODUCTIVE TRACT

The dotted circles indicate the passage of an egg from the right ovary. Sperms are shown entering through the uterus.

of a broomstraw, connect the ovaries with the uterus. The latter organ, also called the womb, is about 3 inches long and shaped like an inverted pear. At the upper end are attached the Fallopian tubes and the neck projects into the upper portion of the vagina. The series of organs together are referred to as the reproductive tract.

Fertilization—the actual merger of sperm and egg—occurs in the narrow confines of a Fallopian tube and the uterus provides a home for the fertilized egg while it grows into an embryonic child.

Ovulation, Menstruation and Estrus—During her lifetime, a normal woman produces more than 300 eggs, usually at the rate of one each 28 days. The discharge of one egg from a follicle which is a sac in the lining of the ovary is known as ovulation.

Only a few of the 300-odd eggs are fertilized. But each time a ripened egg leaves the ovary, nature prepares to protect and nurture the embryo which may ensue. Blood vessels in the uterus fill to the bursting point. When an egg not fertilized in the Fallopian tube passes through, the blood, no longer required, is released. The consequent flow of blood and broken-down cell tissue derived from superficial layers of the lining of the uterus is called menstruation.

Menstruation generally occurs about every 28 days during the reproductive period of a woman's life from puberty to the menopause or "change of life." Each 28-day period is a complete menstrual cycle. The cycle begins with the first day

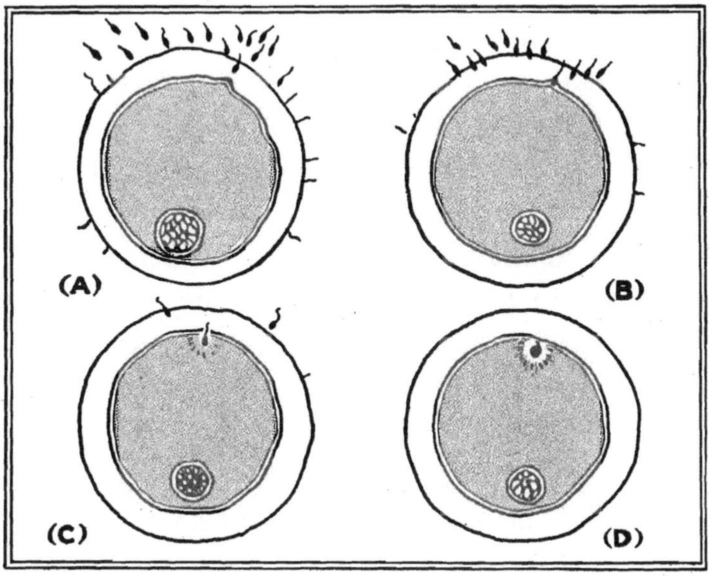

FIGURE 9.—FERTILIZATION OF AN OVUM

(A) A single sperm penetrates. (B) Protective envelope excludes other sperms. (C) Tail of penetrating sperm disappears. (D) Other sperms disappear.

of a menstrual period and lasts to but does not include the first day of the next period.

If the egg fuses with a sperm as it journeys down the oviduct the berth prepared for the embryo in the uterus does not disintegrate. It is needed. Hence, menstruation in this case does not occur and pregnancy ensues. The time of pregnancy between conception and birth is the gestation period.

Menstruation formerly was thought to be the equivalent of estrus—the "heat" or "rut" in lower animals. This idea has now been abandoned for we know that ovulation in humans occurs about the middle of the menstrual cycle. When lower animals are in heat, the fertile female produces her egg and only at this season will receive the male. Humans have no definite mating season and are more or less receptive to sexual congress at any time.

Here it may be pointed out that the 28-day menstrual cycle coincides in length with the lunar month. The moon has no influence upon the phenomenon, although attempts often are made to establish a relationship.

The Race of the Sperms—The male life-reproducing mechanism seems comparatively simple because it has but one job—to manufacture the sperm cells and introduce them into the female. Thereupon the male function ceases.

The sperm factories, corresponding to the ovaries, are two oval glands, the testes. Traveling toward their exit from the male body, the sperms join with the secretions of the prostate, urethral and Cowper's glands to form seminal fluid or semen which is secreted in the seminal vesicles until required in the course of sexual union. Something like 250,000,000 sperms are expelled during the sexual act and each one under a microscope appears to be whipping about vigorously as if eager to be off on its journey toward the egg cell.

For some reason not wholly clear but perhaps similar to the cause of osmosis which lifts sap in a tree, the sperms always propel themselves upstream against the downward direction of the current in the oviduct. Upon their release in the vaginal tract an intense swimming race ensues. Millions of the small but active contestants lose their way and are eliminated, but finally some reach the place where the egg awaits. These survivors now surround the female cell and one darts forward and plunges into it. Fertilization is then accomplished.

The unused sperms, like unfertilized eggs, even-

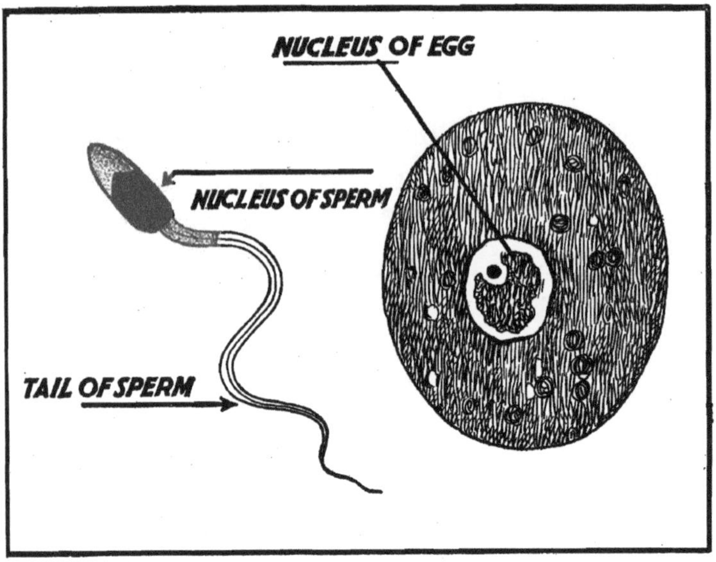

FIGURE 10.—RABBIT SPERM AND OVUM
The rabbit gametes are typical of all mammals.

tually perish. Many ebb out of the vagina, others are killed by vaginal secretions. Those remaining lose their motility and fertilizing power in from 36 to 48 hours and finally die.

All of this is a fully-understood biological process and today it is difficult to believe that sperms first were observed in the seminal fluid in 1677. Not until after 1885 did scientists notice that only one sperm fertilizes the egg.

Bearers of Heredity—It now becomes necessary to look deeper into the interior of the body cells which as we already know consist of nuclei imbedded in cytoplasm.

Within the nucleus appears a swarm of faint grey granules which come together to form a mass of threads, some long, some short. These threads contract into tiny, stubby rods each called a chromosome. Mark the word well, for chromosomes play a vital part in every life. They are the vehicles which carry the mental and physical traits from one generation to another—the literal bearers of hereditary destiny.

Each cell has a definite fixed number of chromosomes and the number is characteristic of the species. The body cell in a human being always has exactly 48, in the mouse 40, and so on.

In the ordinary process of cell division and multiplication the cell tears itself into two parts. Each rod or chromosome splits lengthwise and one-half of it goes into the nucleus of each of the two newly created cells. Thus, each time a human body cell divides during the usual growth procedure there are 48 chromosomes in each of the daughter cells as well as in the one parent cell.

X and Y Chromosomes—We now come to the core of natural sex determination. It is found in the chromosomes.

The human cell which is the forerunner of the ovum contains 24 pairs of identical rods identified as X-chromosomes. But in the cell giving rise to the sperm there are only 23 pairs of X-chromosomes and an unequal pair. One is of the X type while the other, a smaller rod, is labelled Y.

As the forerunner cells form gametes, there is an amazing exception to the usual behavior of the chromosomes. Instead of splitting longitudinally, the whole chromosomes simply divide into two heaps so that the resultant sex cells have only 24 chromosomes each, or one half as many as body cells. It is easy to understand why the sperms and eggs contain only half the full quota of chromosomes, for their purpose is to combine and form one single complete cell of 48 chromosomes—the zygote.

All of the chromosomes in every egg are of the X variety, but there are two sorts of sperms. Fifty per cent of them contain 24 X-chromosomes. The other 50 per cent contain only 23 X-chromosomes

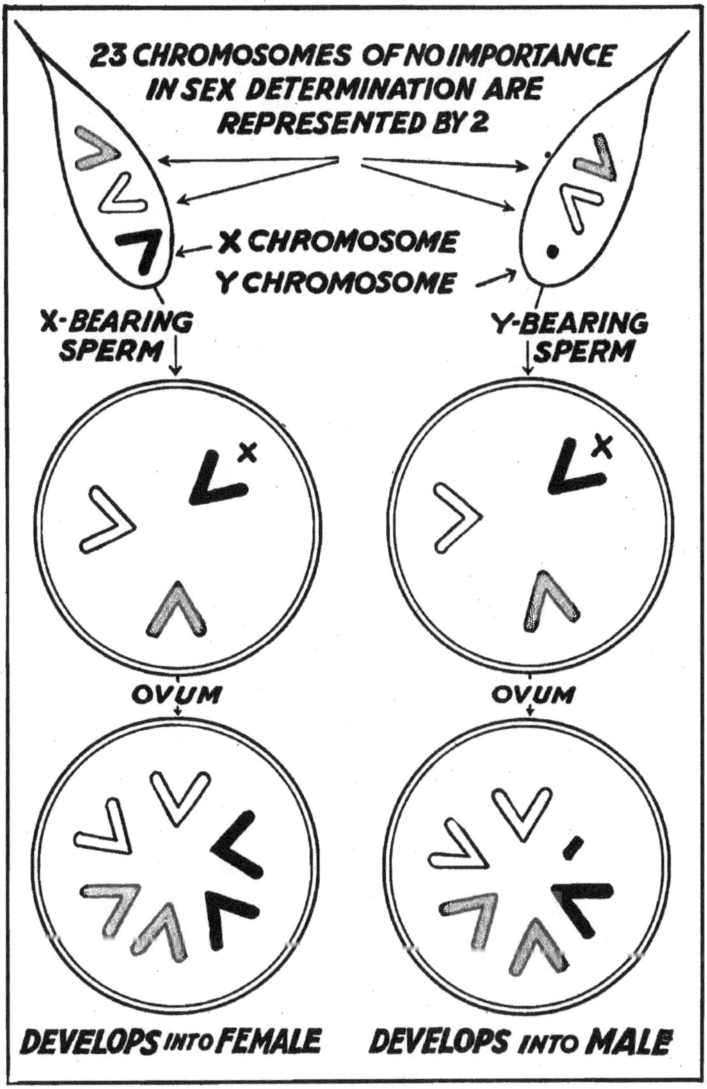

FIGURE 11.—HOW SEX IS DETERMINED

When an X-bearing sperm fuses with an ovum the result is a female. Fertilization by a Y-bearing sperm produces a male.

plus the odd Y-chromosome. On a microscopic slide the sperms bearing Y-chromosomes actually can be distinguished and measured. They have heads smaller by 1/1,000 of a millimeter than those of the X-bearing group.

Here is probably the most important single fact about sex determination. The X-chromosomes are female-begetting and the Y-chromosomes male-begetting. In other words, an egg fertilized with an X-bearing sperm will develop into a girl, while an egg fertilized with a Y-bearing sperm will develop into a boy. The egg has no influence whatever in the matter. The semen from the father alone determines sex. To repeat—all sex chromosomes in the female are exactly alike and the egg consequently is neutral or asexual. Upon the type of sperm which enters it depends the sex of the child.

The same X and Y chromosome rule that applies to man also applies to mice, rats, rabbits, cats, dogs, goats, cattle, horses and other mammals. It also applies to some insects and some—not all—of the lower animals. Since X and Y-bearing sperms are produced by the male sexual mechanism in approximately equal numbers, a 50-50 division of males and females is the result at conception.

While the X and Y chromosome explanation of natural sex determination may seem complex to the lay reader, it is thoroughly established in principle. It stands unchallenged by modern biologists.

How Alcohol Affects Sperms—A number of experiments with parental alcoholism have furnished striking proof that the X and Y-bearing sperms can be influenced artificially.

Leon J. Cole and C. L. Davis, in 1914, demonstrated conclusively that there is a fight or competition for fertilization among sperms. They mated a female rabbit almost simultaneously with two males of different colors. One buck showed itself more potent than the other as the majority of young in the litter always came from it. When the sperms of this buck were damaged by alcoholization, no young were produced by it in the corresponding experiment. They were still produced, however, if the alcoholized buck alone was mated to the doe. The alcoholized sperms thus were capable of fertilization, but were beaten in competition with normal sperms.

Two years later, C. R. Stockard and G. Papanikolau demonstrated that among the progeny of alcoholized fathers the daughters showed a greater tendency to die than the sons. This indicated that definite damage was done to the X-bearing sperms and proved that it is possible, by means of physiological and pathological agents, to influence one kind of sperm—in this case the X-bearing type.

The utmost significance attaches to the latter statement for it lays a firm foundation for artificial sex control.

Scientists in Agreement—In one way or another the ablest men in the field have made it clear that if a way could be found to artifically cause X or Y-bearing sperms to fertilize the egg, sex control would be attained. One wrote several years ago, "It is quite possible that the time will come when geneticists will be able to separate male and female-determining spermatozoa and thus far to a corresponding degree control the sex ratio."

Said another, in 1929: "By mechanical or other methods it may be possible eventually to separate seminal fluid into portions containing predominantly one or the other type of sperm, the X-bearing or the Y-bearing. If this could be done, the control of the sex ratio would be experimentally realizable."

One more may be quoted: "If we could discover some anaesthetic subtle enough to paralyze one kind of sperm and not the other, voluntary sex determination would soon be a practical possibility . . . it would be silly to declare success impossible."

Success was possible and it has been attained. The separating agents or "anaesthetics" have been found, tested and proved effective. They are soda and lactic acid. Soda gives the Y-bearing sperm an advantage in the race to the egg and lactic acid favors the X-bearing sperm. These two substances, therefore, are the master keys to sex control.

III

APPROACHING THE TRUTH

A Variety of Theories—The scientists who have developed the acid-alkali method of sex predetermination by no means brush aside all of the hypotheses and theories dealing with natural determination. On the contrary, they are convinced that several hit very close to the mark and indeed tend to confirm their own conclusions.

No intelligent person, of course, can take any stock in superstitions without any shadow of physical justification, such as the belief that wearing Sunday clothes during coitus affects the sex of the offspring. Such absurdities are not to be confused, however, with either ancient or modern theories based on fact-finding research or even on observation alone.

The theories fall into two major groups: (1) those aimed to regulate sex by discovering and conforming to the natural reproductive processes and (2) those aimed to regulate sex by altering the processes. Those in the first group are discussed in this chapter.

It must be remembered that investigators may discover—indeed may have discovered already—how to produce males or females by acting in accordance with immutable rules of organic behavior. It is likewise true that the acid-alkali method does not eliminate the possibility of finding other and better ways to reach the same goal.

The salient fact is that soda and lactic acid injections into the vaginal tract before mating do predetermine sex. Nevertheless, a study of other theories serves to clarify the acid-alkali method. Since some of these theories are at least partially verified, this chapter is entitled "Approaching the Truth."

Seasons and the Weather—The philosopher Aristotle, living in an age when the Greeks wanted a preponderance of male offspring, advised parents in undertaking physical union to "give preference to the north winds over the south winds."

At first glance the presumption that the elements might influence sex seems preposterous and it probably is of only the slightest, if any, consequence. Down to modern times, however, cold and dry weather has been associated with the conception of males and warm humid weather with the conception of females.

Recently a well-known pathologist said that a study of birth rates and weather reports in two cities revealed that, broadly speaking, "more females are conceived in warm weather and more males in cold weather." He also surmised that more males were born during the glacial eras.

He suggested that as barometric pressure goes up and the temperature down women breathe more oxygen, create more carbon dioxide and thereby change the acid condition of their bodies. Other pathologists challenged his work as not "scientifically satisfying," but if the weather does affect acidity and alkalinity within the female body, the theory cannot be dismissed too lightly.

Passion and Frigidity—It has been noticed since the writing of the Talmud that amorousness and intensity of passion in a woman are likely to be followed by the birth of a boy, whereas when she is unresponsive and frigid a girl is more often conceived.

In the Tracte Berakoth, much quoted by medical

Talmudists, is the passage: "If the husband bring forth semen first then she will bear a female, if the wife bring semen first then she will bear a male." King Louis Philippe of France, in the 19th century, amazed his court by accurately predicting the sex of his children. He explained, "Quand je veux une fille, je l'offre á la reine: quand je veux un garçon, j'attends qu'elle me l'offre." Roughly translated this means, "When I want a girl I offer it to the queen: when I want a boy I wait for her to offer it to me." Otherwise stated, if the woman experienced the orgasm first, she conceived a boy; if the man experienced the orgasm first, she conceived a girl.

During an African tribal war, described in the *Edinburgh Medical Journal* of August, 1886, the victors killed all the enemy males and took the females as wives. The captive women gave birth to 403 girls and 79 boys. Under the tragic circumstances attending conception it may be taken as likely that orgasms failed to occur in a majority of the prisoner-wives.

We know now that these observations and events may rest on a sound scientific foundation. The cervix or neck of the womb secretes an alkaline fluid. If the woman is sufficiently aroused to a complete and strong orgasm, this cervical secretion is freely liberated. It is altogether likely that the resulting alkalinity of the vagina encourages the Y-chromosome bearing sperm to effect conception.

Such an assumption dovetails perfectly with the acid-alkali method whereby the pH or acid-alkali balance of the vaginal tract is artificially altered to control sex.

Ages of Father and Mother—Fascinating, but contradictory and inconclusive, are the beliefs that the ages of the parents have something to do with the sex of the child.

One school of thought reasons that the older parent stamps its own sex upon the offspring—that seniority on the father's side, for example, results in more boys than girls. The other school holds that where there is a marked disparity in ages, the younger parent induces its own sex in the offspring. Obviously one, if not both, of these hypotheses is wrong.

Yet no less an authority than Dr. Harold G. Taylor, a pioneer in developing the acid-alkali method, has seriously cited the idea that union of a young wife and a middle-aged husband will most likely produce a girl and union of an older woman and a young man, a boy. Dr. Taylor declared, "It stands to reason that a man of 30 will be more active and passionate in coitus than an inexperienced bride who may experience no orgasm at all. If both are of equal ages the chances are that a boy will result, as age for age the woman should be the better animal sexually. When the young husband has grown more sexually inclined or the wife has become less interested, then he probably will produce daughters."

Thus, if feminine passion alters the acid-alkali balance in the vagina and if age affects passion, then age may affect the sex of the child. Experiments with rats have indicated that this may be true in some degree.

Parental Vigor or Weakness—Keeping in mind the possibility that a quicker or stronger orgasm by one parent may tend to produce a child of the opposite sex, now consider the allied possibility that the physical condition of the parents figures in the procedure.

A book entitled *The Law of Sex* written by George B. Starkweather in 1883 give this idea wide publicity. Starkweather stated that "the superior parent produces the opposite sex," adding that "superiority means anything that tends to increase functional energy in any part of the system."

The hypothesis that the sex of the young tends to be that of the weaker parent was tried in 30-odd mating experiments with Texas cattle. Poorly-fed bulls in a weakened sexual condition were mated with cows in prime condition and were reported to have produced an excess of bull calves. When strong bulls were mated to weakened cows, the breeders claimed an excess of heifer calves.

Because of the physiological vagueness of "superiority" and the lack of more extensive experiments, the vigor-weakness theory is hardly suitable for scientific discussion although, due to the possible link between response during coitus and the pH or acid-alkali balance of the vagina, it warrants consideration.

Left and Right Ovaries—We now come to a belief wholly without foundation. It is described here only because of its widespread popularity and curious tenacity in the face of positive contradictory evidence.

The Greek philosopher Anaxagoras thought

that the right ovary always produced male children and the left ovary females. An Englishman, E. Rumley Dawson, revived the ancient idea some years ago and supported it by maintaining that the female sex, being the weaker, would "naturally" come from the left side of the body, this being the "weaker" of the two. Other writers copied Rumley and down to the present day wives sleep on their right sides in the hope that a boy will result!

In 1922 an English inventor actually patented a "sex calculator" based in part on "the well-known" theory that "one ovary always produces male ova and the other female ova."

The absurdity of such a notion is at once apparent. In the first place, the ovaries and ova have nothing to do with sex determination. The ova are asexual or neutral and the sperm is the exclusive deciding factor. Moreover, obstetricians have recorded many cases in which children of both sexes have come from the eggs of one ovary, the other ovary having been removed by surgery.

Speaking of the Rumley fallacy, Doctor Taylor observed ironically that if correct, "a national disappointment could be prevented by removing the queen's left ovary." It may be added that wives sleeping on their left sides to beget boys might as sensibly sleep on their heads!

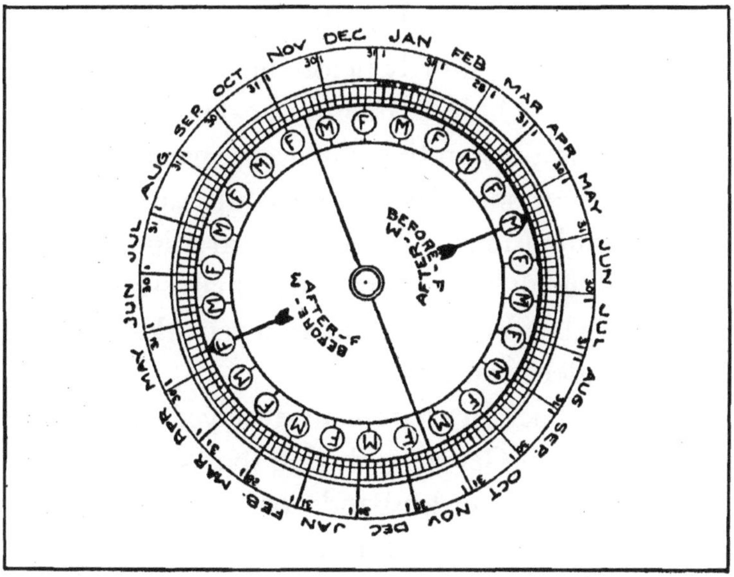

FIGURE 12.—PATENTED "SEX CALCULATOR"

This device, based on the erroneous right and left ovary theory, was patented in 1920.

The Menstrual Cycle—At once the most widely-held, complicated and exciting theory of sex control without the use of outside substances originated with Hippocrates and now seems just as logical as the right-and-left ovary theory is illogical.

From careful observation Hippocrates concluded that conception shortly after a woman's menstrual period would produce a son, while conception occurring shortly before her period would result in a daughter. Broadly speaking, this was probably the outstanding discovery made during 2,000 years of speculation and conjecture about sex determination. Although variations still make it difficult to fit it to individuals, the basic principle has been borne out to a remarkable extent by modern research.

The Hippocratic statement received startling confirmation in Germany during the World War. Because the soldier-husbands of many German women visited them for only 48-hour leave periods, it was easy to gauge at what time conception took place in relation to menstruation. P. W. Seigel kept records and reported his findings. According to his data, coitus during the first days after the cessation of menstruation resulted in a preponderance of male births, whereas coitus between ovulation and the following menstruation gave a preponderance of females.

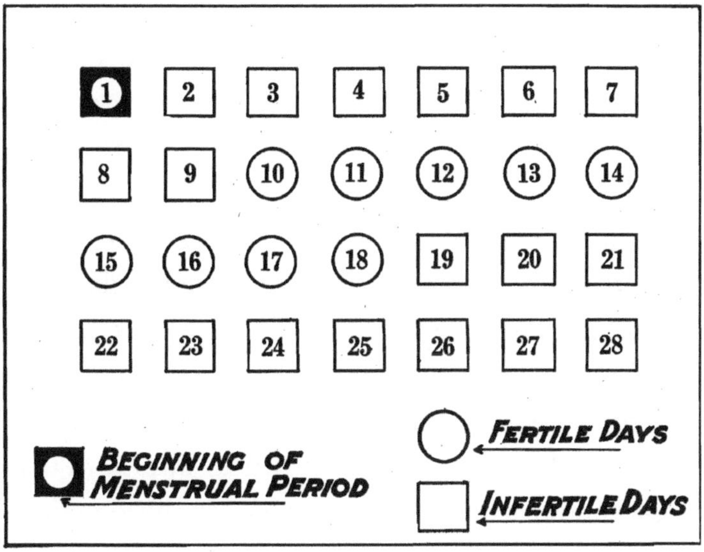

FIGURE 13.—A MENSTRUAL PERIOD CALENDAR

The figures apply to the menstrual month, beginning with the first day of menstruation.

How can this apparent phenomenon be explained?

Doctor Taylor and his brother, Dr. Gordon Kirwin-Taylor, after testing the pH in many women, found that "vaginal secretion is usually nearly neutral immediately after a period, and as a rule becomes progressively more acid until the next period."

Here, then, is a satisfying explanation of the

Seigel data. The secretion is more apt to be alkaline immediately after the period and more acid later in the cycle. Therefore, more boys were produced by the earlier conceptions and more girls from the later conceptions.

Variations in the Cycle—It would logically follow that the rhythm of the menstrual cycle provides a full and complete answer to the sex determination puzzle—if you desire a male child secure intercourse as soon after menstruation as possible and for a female child wait until later. This simple theory will not always work for three major reasons: (1) there is a wide range of variations from the average 28-day period; (2) the acid or alkali balance may be naturally altered during the period by secretions of varous organs within the vaginal tract; (3) individual chronic conditions may affect the pH.

Few women menstruate at exactly 28-day intervals. They may be a few days early or late. Actual experience shows that such variations are quite normal and the cycle may be from 21 to 35 days in perfectly healthy women. Furthermore, the length of the cycle varies in the same woman from time to time, depending on external factors including climate, illness and emotional upsets.

We already have seen how the pH of the vagina

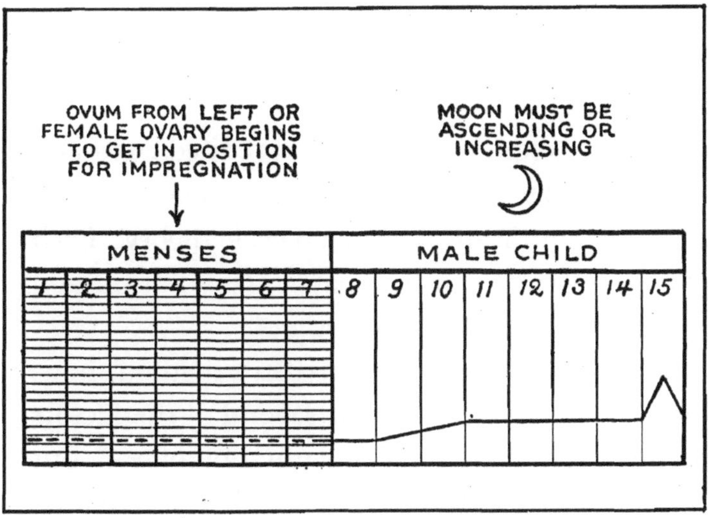

FIGURE 14.—LUNAR THEORY OF SEX CONTROL

A section of a menstrual cycle chart based partly on notions that the ovaries and the moon influence sex.

may be temporarily changed at the moment of orgasm by the alkaline cervical secretion. The variation caused by this and other glandular secretions again makes the determination by the menstrual cycle uncertain.

Stepping Stones to Success—We now have reviewed each of the major theories which cast any light on essential factors in determining sex. There exist combinations of these ideas too numerous to mention, and "new" theories usually turn out to be two or more of the old ones scrambled together.

As often as not, the utterly discredited right-and left ovary fallacy is found in conjunction with the extremely significant menstrual cycle theory. Each theorist is wedded to his particular solution and inclined to be intolerant of others.

The developers of the acid-alkali method, however, do not deny that one or several of the factors discussed bear upon the determination of sex. But it should be emphasized again that the efforts under consideration so far have been directed toward the regulation of sex by analyzing natural influences upon determination. It is wholly reasonable to assert that these efforts to some extent have been successful. Certainly they help to explain why the acid-alkali procedure is successful.

The distinction is largely one of hypotheses and theories as against method—ways of adjustment as contrasted to actual control. Whatever the natural influences which determine sex, they are overcome by the acid-alkali technique as proved in actual practice.

IV

PIONEERING IN GERMANY

The Scientific Groundwork—The acid-alkali method of sex control as we have seen rests on a sound fundamental of modern knowledge—the unassailable X and Y chromosome theory accepted by all modern biologists. This already has been outlined, but a restatement of the basic facts will clarify the story of actual experimental work to follow.

There are two kinds of sperms, X-bearing or female-begetting and Y-bearing or male-begetting. If an X-bearing sperm fertilizes the egg, a female results; if a Y-bearing sperm does the fertilizing, a male results. This law is fixed, uniform and invariable. There are no exceptions.

The acid-alkali method is the only known way to artificially affect the sperms so that the type arbitrarily selected will fertilize the egg. Alkalinity of the medium or fluid through which the sperms reach the egg will bring about fertilization by a Y-bearing sperm; acidity of the medium will bring about fertilization by an X-bearing sperm.

The pH or acid-alkali balance of the vaginal fluid varies in different species under certain circumstances and during certain periods. But it can be raised or lowered at discretion by the use of acid or alkaline substances.

Soda, an alkali, and lactic acid are the chemical agents so far found most effective for this purpose and their direct introduction into the vaginal tract in solution, powder or tablet form just before coitus is the most satisfactory means of applying them.

Keeping these fundamentals in mind, let us scan the chronological advances which led to every-day use of the method by thousands of animal breeders and the probability of widespread adaptation to human beings in the near future.

The Father of Sex Control—It is axiomatic that many of the greatest discoveries of science are made accidentally, revealed quietly, perfected slowly and utilized cautiously. The history of true sex control follows the traditional pattern.

On February 21, 1930, there appeared in the

German Medical Weekly a modest article by Prof. Felix Unterberger, chief of the Gynaecological Department of Mercy Hospital, Koenigsberg, in eastern Prussia. It was entitled "The Problem of Planned Sex Determination in Man." Two years later in 1932 the *Medical Weekly* printed another Unterberger article headed "Hydrogen Concentration and Sex Determination" and the *German Medical World* printed a third, "Chemical Influences Upon Ova and Spermatozoa."

These three articles for the first time revealed the solution of the ancient sex control riddle and marked Unterberger as father of the new science. Far-reaching improvements followed, but Unterberger hit upon the essential formula.

His disclosures caused a minor uproar in the German press. Newspapers elsewhere ignored them, although on March 29, 1930, the *Journal of the American Medical Association* carried a brief resume of the original article from its Berlin correspondent.

Despite later confirmation of the Unterberger thesis by two English doctors, Americans, except for a few geneticists, remained unaware of the German's achievements before the momentous experiments sponsored by *The New York Daily News* in 1938 and 1939.

Soda and Sterility—For generations in many lands veterinarians and breeders have known that soda rinsings or washings help to overcome sterility or failure to conceive in dogs, cows and horses. An ordinary treatment for barrenness consists of a vaginal lavage with a tablespoonful of soda in a quart of water. This old-fashioned remedy led Unterberger to investigate the acidity of vaginal contents in women.

"I found that those among my patients who had consulted me regarding sterility showed a surprisingly high rate of acid in the vagina," the German reported. He continued:

"Now it is commonly known that washings with sodium bicarbonate prior to cohabitation is completely harmless. Consequently, I failed to see why we doctors should not apply this harmless means against sterility in cases where an abnormally high degree of vaginal acid was diagnosed. And whenever there were no other symptoms of disease either in the man or woman, the result of this simple means was often surprising.

"The much desired pregnancy often occurred

already after the next monthly period and the child was—invariably—a boy. This made me suspicious. Could it be that the chemical status of the vagina exercised an influence over the male spermatozoa?"

It could and did exercise a most profound influence, as Unterberger and his successors have proved over and over again.

First Trials With Humans—The Prussian professor eagerly studied everything about vaginal secretions on which he could lay his hands. It previously had been established that if tested with litmus paper, the vaginal bacterial flora or vegetable microorganisms normally are acid. Further, it was known that this acid condition varied during the menstrual cycle and finally that it could be temporarily altered by alkaline secretions from the Fallopian tubes, uterus and cervix.

It also was known that male spermatic fluid is ordinarily slightly alkaline, although it too is variable depending upon gland secretions passing into it.

Assuming from his observations that a strongly alkaline condition at conception favored the birth of boys, Unterberger reasoned that a strongly acid condition would favor the birth of girls. He decided to put this hypothesis to the test. In some cases Unterberger instructed husbands to sprinkle powdered soda sparingly on the prepuce and glans just before intercourse; in others, he suggested a soda solution douche by the woman.

Out of a series of 54 conceptions, Unterberger thereby produced 53 boys and 1 girl. The single failure resulted when the husband did not use the prescription because his wife already was pregnant. Announcing his success, Unterberger wrote:

"My experiments brought no less than 100 per cent boys. Now, I am not so blind as to believe that now we have a means to determine the sex of an unborn child with 100 per cent certainty. But, even if we succeed in only 90 per cent, the result would surely be a remarkable one."

Further Unterberger Successes—After an interval of two years Unterberger in his 1932 articles reported fresh results and explained them in the light of the X and Y chromosome theory.

He chose for his second series of human experiments 20 wives who previously had borne from 1 to 4 girl babies each. By the use of his method,

19 out of 20 gave birth to boys and he could not learn whether his orders were observed in the twentieth case. Thus, in the end, he had obtained a total of 73 boys from 74 conceptions.

Since all of the German patients wanted boys, Unterberger did not try to obtain girls, but he made this prophetic observation:

"Unfortunately I am not in a position to advise how to favor the birth of girls. I still lack practical experience. Nevertheless, theoretical deliberation should not exclude the mere possibility. An increased degree of vaginal acid, caused perhaps by strongly diluted lactic acid solution, should in my estimation highly favor the procreation of girls."

Reactions in Other Countries—The 1932 Unterberger papers attracted more scientific attention than those of 1930, but the reactions outside of Germany were somewhat critical. British scientists, unaware that the Taylor brothers in London already were deep in research work, were cool as was the *Journal of the American Medical Association.*

The A. M. A. magazine, on February 18, 1933, commented: "One is entitled to a prolonged healthy skepticism in this matter. Furthermore, if one admits that an alkali can change such a momentous biologic action as the act of fertilization, one would logically fear the possibilities of other histologic or pathologic, perhaps catastrophic, changes in the resultant product of conception."

Not so quick to dismiss the subject were Prof. Leon J. Cole and Prof. Ivar Johansson, nationally known geneticists of the University of Wisconsin. In the *Journal of Heredity* for July, 1933, they questioned the reliability of the reports from Unterberger's patients, but added this forthright assertion:

"If Unterberger's records of an unbroken series of 74 male births as a result of treatment with sodium bicarbonate could be considered as a random-sample of treated individuals, the positive effect of the treatment would be practically certain. The probability of obtaining such a result where chance only is operating would be 0.5 or practically zero."

The Fueth Tests—Nine months after Unterberger's first epochal announcement, H. Fueth, a Cologne professor, published in the *Munich Medical Weekly Journal* of November 31, 1930, an

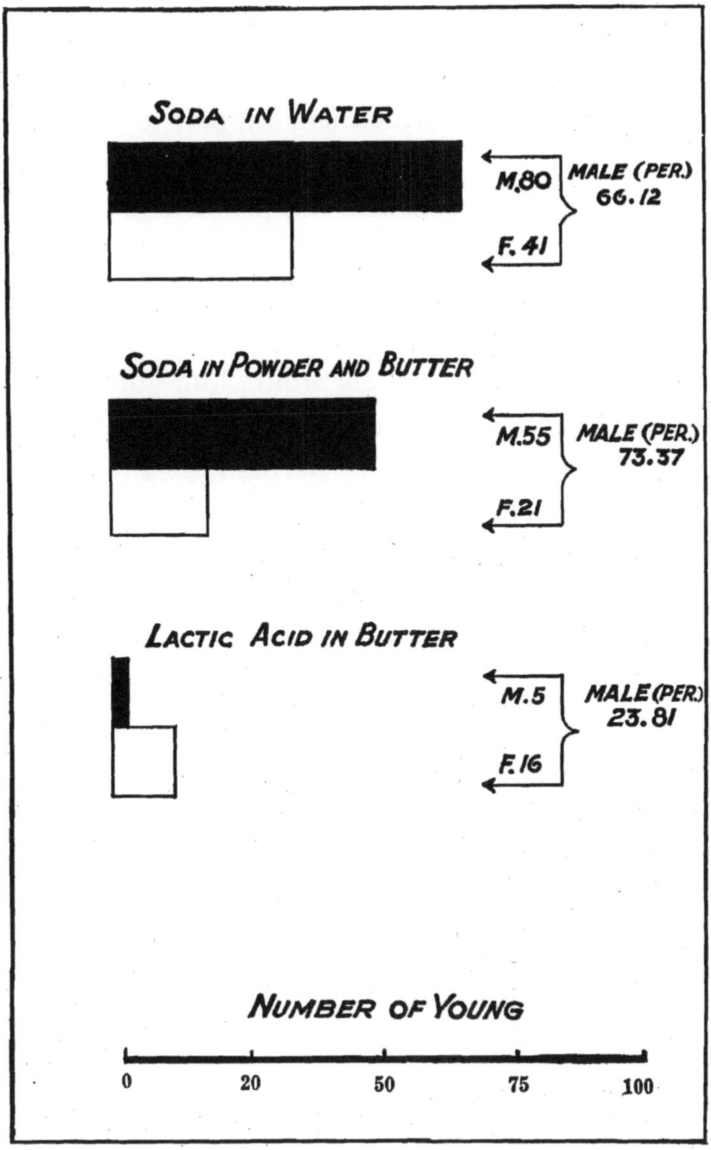

FIGURE 15.—RESULTS OF RABBIT TESTS

This chart shows the remarkable outcome of the Unterberger-Kirsch experiments with soda and lactic acid.

article headed "Question of Arbitrary Douching in Human Beings" which lent support to the acid-alkali theory.

Fueth had approached the problem in a new way. He investigated the "purity" or bacterial content of the vaginal secretions of 253 women during the final month of pregnancy. Many of these expectant mothers told him they had douched with soda solutions after reading of Unterberger's patients in the German press.

Fueth found that the women with acid secretions gave birth to a preponderance of girls, while those with alkaline reactions produced a majority of boys. These independent observations strongly corroborated Unterberger, indicating again that the female-begetting sperms prefer an acid medium and the male-begetting sperms an alkaline medium.

An Experiment With Rabbits—With an associate, W. Kirsch, Unterberger conducted a brief series of tests in 1932 on rabbits which, while statistically inconclusive, forecast the mass results with rats and rabbits obtained 6 years later in America.

The Germans tried three treatments: (1) vaginal washings with 0.25 per cent soda solutions immediately before coitus, (2) administration of soda as powder or mixed with cacao butter, and (3) administration of lactic acid in cacao butter. The following results were obtained:

	No. of Young	*M*	*F*	*Pct. Male*
Soda in water	121	80	41	66.12
Soda in powder and butter	76	55	21	73.37
Lactic acid in butter	21	5	16	23.81

There were too few of these animal tests, especially with the acid, to be of numerical significance, but as far as they went the results were sensational. They excited the curiosity of later investigators and finally pointed the way toward the triumph of *The News* with experimental animals in the United States.

V

PROGRESS IN ENGLAND

The British Awakening—The staid English medical world was shaken from top to bottom in 1934 when it found the sex control work of a comparatively unknown German, Unterberger, laid upon its doorstep with the confirmation of two eminent young physicians and consultant specialists at various London hospitals.

Dr. Harold G. Taylor and his brother, Dr. Gordon Kirwin-Taylor, who studied together at Cambridge University had been interested in predetermination for several years and carried on their research as gynaecological surgeons at St. George's Hospital. They thoroughly examined and tested the German theory. On February 24, 1934, they described their work in a British medical magazine, *The Lancet*.

The essay revealed that the Taylors had begun where Unterberger and Fueth finished and on top of substantiating the German discoveries generally, had added extremely important original material regarding the influence of the chemical reaction of the vaginal secretion at the time of coitus on the sex of the resulting child.

A pertinent paragraph in *The Lancet* article showed that the Taylors held the conventional view of natural sex determination. They wrote:

"It is now generally agreed that in the human species, the spermatozoa are of two kinds—differing in the number of their chromosomes. One variety is believed to be potentially male, the other female. The human female ova are, of themselves, asexual. If an ovum is fertilized by a 'male' spermatozoon, the resulting child is a son—if by a 'female' spermatozoon, it is a daughter. Work done in Germany and elsewhere seems to show that a relatively alkaline secretion favours fertilization by the male spermatozoa, while an acid secretion favours fertilization by the female, and this opens up the possibility of pre-arranging the sex of the offspring."

First Taylor Observations—With typical British thoroughness and perseverance the Taylors began to examine women patients and tally their records. At the start they examined women at frequent intervals over a period of several months, but thinking the vaginal secretion reactions of an individual might vary during the menstrual cycle, they made more tests at different stages of the cycle. They also noted the chemical ingredients in cleansing douches and contraceptives used by the patients.

Material for the tests was collected from the vagina on a camel-hair brush and its pH determined by a micro-colorimetric method. Usually the reaction of the cervical secretion also was tested. The Taylors reported five "tentative conclusions" which may be summarized as follows:

(1) Variations in the pH of the vaginal tract occur, in most cases, during the intramenstrual period. The cervical variations are small and do not always parallel the vaginal variations.

(2) The vaginal secretion usually is neutral immediately after a period and becomes progressively more acid until the next period, although in exceptional cases the acidity is continuous.

(3) It is possible that variations in one person tend to be constant under uniform conditions.

(4) It is likely that the variations are associated with different vaginal bacteria.

(5) The vaginal reaction can be modified by local treatment, and, owing to a flow of alkaline secretion from the cervix, by sexual relations.

These "tentative conclusions" clarified what had been strongly suspected by Unterberger and hinted at in the speculations of natural determination investigators.

Further Facts About the pH—The Taylors steadfastly continued their work in England and Dr. Harold G. Taylor announced further progress in a lecture to post-graduates at St. George's Hospital on October 1, 1937.

"We are of the opinion," he said, "that a vaginal pH of above 7 is favorable to boy conception and a pH below 6 to girl conception, but we like to get a pH nearer to 8 before allowing a son to be attempted and a pH about 5.5 for a girl.

"We may find in this particular patient we are advising that the pH of her vaginal and cervical secretions conforms for a time at least with what is our idea for the desired sex, and then perhaps that it rapidly changes. In such a case we would advise her making the attempt at conception on those days when her secretion is favorable. In order to be doubly safe we might find it advisable

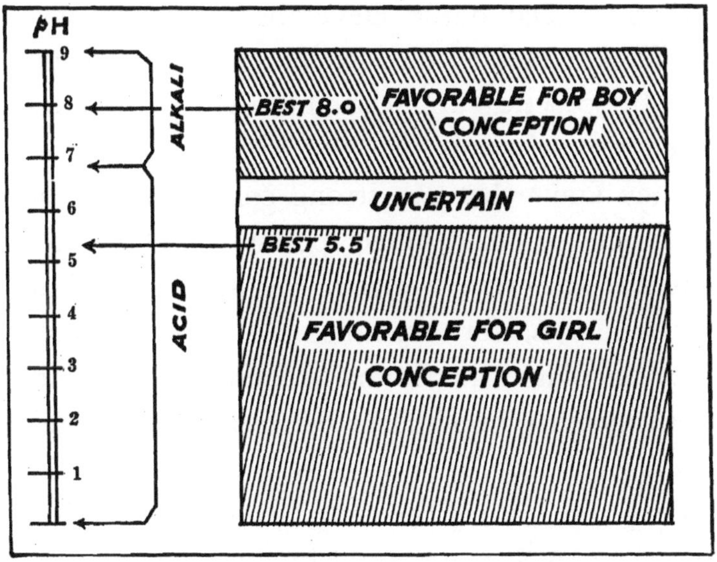

FIGURE 16.—A pH CHART FOR WOMEN

The shaded areas and arrows show the proper acid-alkali balances for sex determination as found by the Taylors of London.

to order her an alkaline douche to be given some hours before coitus so as to be quite sure of the alkalinity.

"From time to time we find that at no time is the secretion of the pH right for the desired child, and in such cases one may have to effect a complete change in the pH of the vaginal secretion by giving stronger treatment at a time nearer coitus."

Human Technique Developed—One by one the Taylors tackled the specific problems arising from the practical application of their pH studies to feminine patients.

They knew that too strong concentrations of either acid or alkali killed sperms and often prevented conception altogether. Most chemical contraceptives, for example, are strongly acid. They calculated that human sperms can live and retain motility in a pH of between about 4 and 10 and, therefore, kept it between those extremes.

They also studied the time factor to determine just when treatment should be given in relation to the natural pH and coitus and also tested various specific chemicals and types of douches, always striving toward an ideal technique.

From the start they found the bicarbonate of soda douche to be fairly successful for boy conception, but in 1937 they were trying douches of di-sodium hydrogen phosphate. To encourage girl conception, they began by using boric acid but later changed to lactic acid spuman tablets which are effervescent and thoroughly acidify the surface membranes. They also produced acid reactions with the mono-sodium di-hydrogen phosphate.

Seeking an absolutely reliable syringe, they agreed that a fully effective "whirling spray" type would be best because it allowed a small quanity of fluid—3 to 4 ounces—to be injected and reinjected several times, at the same time reaching every crevice of the vaginal tract.

When using a solution, they were careful to see that it was of approximately the same temperature as the tissue fluids. A solution too warm, they reasoned, would cause a flow of secretion from the tissues which might upset the expected results.

Taylor Results With Women—The ultimate worth of any scientific theory or method must be estimated from the results in actual experiments. Does it really work?

By 1937 the Taylors were able to announce that up to that time they had achieved success in 73 per cent of their cases. A year later they lifted the percentage of success to 84, despite the hazards of human error and carelessness in following instructions when a larger number of patients were involved.

In his 1937 lecture Dr. Harold G. Taylor

modestly said: "It may be said that many magicians have made their reputations on the even chance. Our results, gentlemen, have been better than that, and definitely we are not magicians. We obviously cannot guarantee results, but we believe with Unterberger that we may be on the threshold of a big discovery."

Use of Force and Electricity—It was inevitable upon the wide acceptance of the male-bearing and female-bearing sperm theory that efforts would be made to separate the sperms by mechanical, physical and dietary means as well as by the direct chemical application method of Unterberger and the Taylors.

Theoretically it is altogether possible that alternate ways of achieving the separation and thereby controlling sex may be discovered. For this reason several such attempts are noted here before advancing to the third stage of the chemical method development in the United States which swept away lingering doubts and placed the acid-alkali procedure at the disposal of practical animal breeders.

The theory that the Y-bearing sperm carries a positive electric charge and the X-bearing sperm

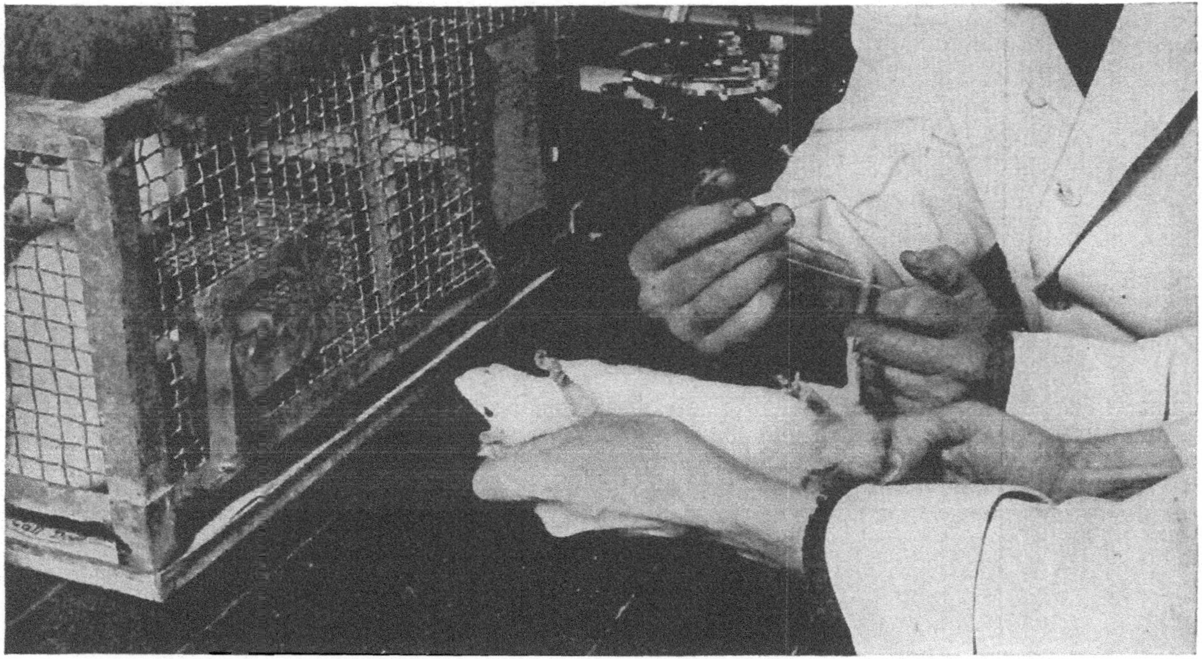

FIGURE 17.—STARTING A pH TEST

Laboratory workers taking a smear for use in determining the acid-alkali balance of the vaginal tract of a rat.

a negative charge was advanced in 1933 by two Russian biologists, N. K. Koltzoff and V. N. Schroeder.

They passed an electrical current through a rabbit sperm suspension and then artificially inseminated a doe with sperms from the anode. She produced 6 young, all females. Another doe, impregnated with sperms from the cathode, produced 4 males and 1 female. A third doe was impregnated with sperms from the central part of the tube and gave birth to 2 males and 2 females.

While these results were startling, three impregnations are virtually meaningless to scientists. Later Dr. Fred E. D'Armour, University of Denver endocrinologist, failed to produce offspring after using a similar apparatus.

J. L. Lush, in 1925, on the strength of a possible difference in the specific gravity of the two kinds of spermatozoa in the rabbit, attempted to separate them by centrifugating and then artificially impregnating unbred females with the different factions. His results were negative.

The possibility that centrifugating may have some effect on the chromosomes, however, has been indicated by Dr. H. W. Bean and Dr. R. L.

FIGURE 18.—FINDING THE ACID-ALKALI BALANCE

A biologist at the Applied Research Laboratories completing a pH test.

King of Princeton University. Whirling in an ultra-centrifuge until a force of 100,000 times gravity is generated, they reported, will sometimes arrest cell division and produce cells with 2 or 3 times the normal number of chromosomes.

Sex Control and Feeding—Because remedies from time immemorial have been taken through the mouth, numerous investigators have tried to influence the sex of offspring through special foods for the parents. These attempts have not as yet proved effective although again it must be said that the method seems, and sometime may be proved, feasible.

Doctor D'Armour during 1933 tried mixing soda and lactic acid with food to influence the sex of rats. He placed the animals on a stock diet plus 2 to 5.5 per cent sodium bicarbonate and 2.5 to 5 per cent lactic acid. He publshed his results in the magazine *Science,* January 19, 1934.

The soda-fed animals produced a total of 168 females and 146 males, a sex ratio of 1.15, while the acid-fed animals produced 170 females and 145 males, a ratio of 1.17. In 14 control litters the sex ratio of females to males was 1.03. D'Armour concluded that there was no physiological basis for the belief that feeding influenced sex.

Similar negative results in feeding experiments were obtained in preliminary work under sponsorship of *The News* which then tested the douching method with sensational success.

VI

TRIUMPH IN AMERICA

The Role of the News—The curiosity, courage and influence of a great American newspaper, *The New York Daily News,* lifted the science of sex control out of obscurity and proclaimed it to more than 3,000,000 readers as a sound and practical reality. Up to this point the history of the acid-alkali method falls roughly into three periods in three countries:

(1) Discovery—Unterberger in Germany.

(2) Progress—The Taylors in England.

(3) Final Proof—*The News* in America.

Newspapers frequently have provided the impetus and resources for trail-blazing explorations and scientific research. The memorable journey of Sir Henry M. Stanley into Africa to find Dr. David Livingstone comes to mind as an example. Stanley's expedition, which opened the way for exploration of the dark continent, was sponsored by

James Gordon Bennett, proprietor of *The New York Herald.*

J. M. Patterson, president of *The News,* which in 1937 was cited by the Pulitzer Prize Committee for an unparalleled campaign against venereal diseases, learned of the European sex predetermination work early in 1938. He initiated inquiries at home and his news gatherers reported that the United States with all its laboratories and research foundations had contributed little, if anything, to the development of something with tremendous potentialities.

Patterson decided to go straight to the heart of the matter and cast about among America's leading geneticists to find a suitable experimental agency. Several recommended the Applied Research Laboratories at Dayton, N. J., a commercial institution directed by Thomas Durfee and approved as a public health laboratory by the State of New Jersey.

Durfee, a biologist formerly assigned to the Rockefeller Institute, undertook with the assistance of his laboratory staff the first of a series of mass experiments for *The News* on February 9,

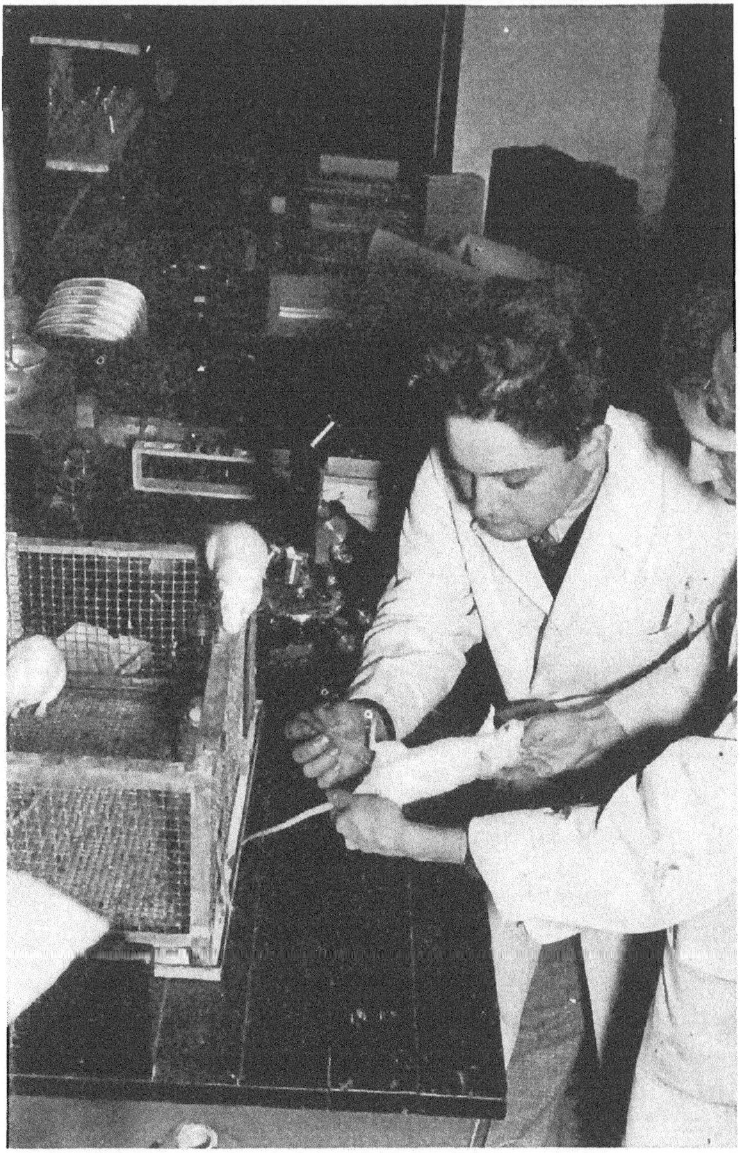

FIGURE 19.—A SEX CONTROL TREATMENT

Laboratory workers douche a female rat with alkaline solution to produce male offspring.

1938, which led to brilliant achievements of the first magnitude.

Preliminary Diet Experiment—With the caution and skepticism typical of scientific realists, Durfee refused "to predict or even hazard a guess as to the outcome to such an investigation." The publisher, too, remained open-minded but resolved at the risk of ridicule and failure to put the European theory on trial and obtain a decisive verdict one way or the other. He agreed with Durfee that "the study to be of value must be most carefully and scrupulously conducted."

By mutual consent it was decided to use white rats from a colony of 5,000 maintained for general scientific purposes as experimental animals and to begin by attempting to influence sex by the ingestation process which might offer advantages in simplicity and economy. This experiment, like all other diet tests so far, proved futile.

From a group of females each with laboratory records of 2 previous litters, 20 rats were selected at random. The sex ratio for their first 20 litters had been 106.7 males to 100 females and for their second 20 litters 109.9 to 100, in close conformity to the accepted rat ratio of 105-106 males to

100 females. The preponderantly male-bearing mothers were acid-fed and the preponderantly female-bearing mothers alkali-fed. The previous litter sex ratio of the acid-fed group was 171.4 males to 100 females, and of the alkali-fed group 68.5 males to 100 females.

For two weeks before mating 5 per cent by weight of lactic acid or 5 per cent bicarbonate of soda was mixed with the regular colony diet in the cages of the animals undergoing the test. The sires were continued on the basal colony diet. Seven litters were produced in each group as follows:

Acid-fed Group			*Alkali-fed Group*		
Litter No.	M	F	Litter No.	M	F
1	5	4	1	5	6
2	6	4	2	1	6
3	3	3	3	5	5
4	7	1	4	3	6
5	1	1	5	3	3
6	4	5	6	4	5
7	4	3	7	4	5
Totals	30	19	Totals	26	36

The sex ratio of the young born to the acid-fed mothers turned out to be 157.9 males to 100 females and of the young born to the soda-fed mothers 72.2 males to 100 females. Durfee on July 11, 1938, reported to *The News:*

"No significant differences in the sex ratios of offspring from either group could be demonstrated as a result of the treatments when compared to similar data for previous litters. Therefore, we conclude that the pH of the vaginal tract in rats insofar as it affects the sex ratio cannot be altered by dietary measures tested in this experiment."

So ended the diet experiment. But much had been learned—so much that tests of the direct douching method were ordered immediately. The three subsequent douching series, for convenience, were called Rat Experiment No. 1, Rat Experiment No. 2 and Rabbit Experiment.

The pH of Rats—Following the English procedure Durfee, prior to the feeding tests, had taken vaginal smears daily from each rat for a period of ten days. These he tested colorimetrically for their pH reactions and the stage of estrus.

These tests indicated a definite correlation between the pH and the estrus cycle but a lack of uniformity between individuals. Durfee also found a slight tendency in the female-bearing group toward greater acidity and in the male-bearing group toward greater alkalinity. This coincided perfectly with the observations of Unterberger, Fueth and the Taylors. Durfee concluded that direct vaginal douching would materially alter the pH where feeding had failed to do so.

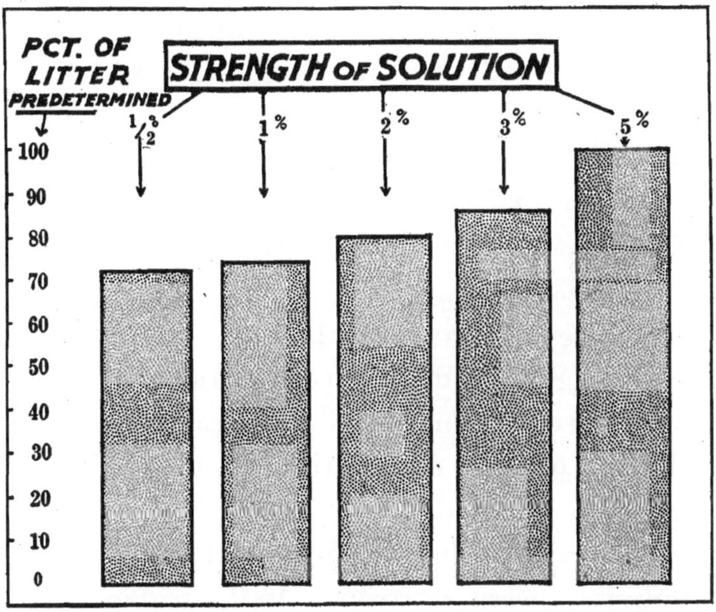

FIGURE 20.—STRONGER SOLUTIONS MORE EFFECTIVE

The chart shows how the desired sex preponderances rose as solution strengths were increased in Rat Experiment No. 1.

Meanwhile he had carefully studied an unsuccessful attempt by D'Armour in 1934-35 to apply acid and alkali by means of saturated cotton vaginal plugs. He reasoned that solutions used to saturate the plug were too weak to counteract the normal vaginal secretions, that the semen may have been ejected above the plug and that no definite precautions were taken to secure prompt matings. With the D'Armour failure in mind, he devised a procedure to safeguard against such errors.

Durfee believed the success of the work depended wholly upon assurance that sufficient amounts of the solutions reached the vaginal tract and that conception should occur before their effects could be neutralized by the natural secretions of the vagina.

Technique for Rats—With little information available to guide him, Durfee empirically chose to mix 5 per cent solutions of pure reagent grade bicarbonate of soda and lactic acid in distilled water for the initial douching treatments.

He ascertained almost immediately that the concentrations were too strong. In the acid group no conceptions at all were obtained. With the 5 per cent soda solution, 2 litters were obtained

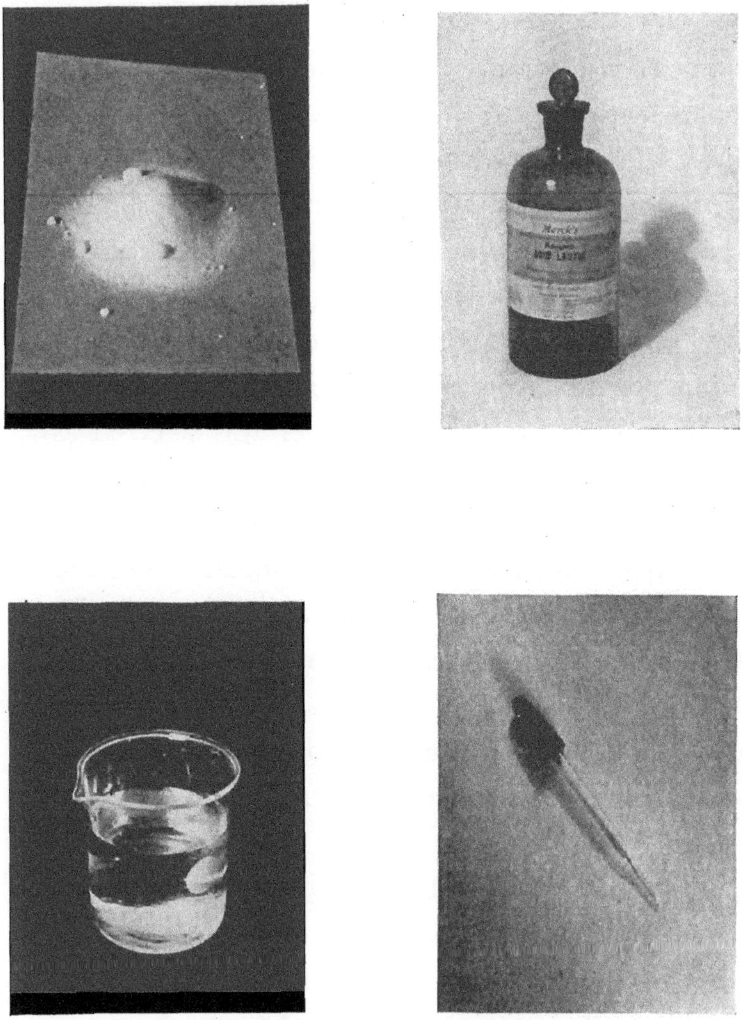

FIGURE 21.—EQUIPMENT FOR RATS

Ingredients and apparatus used at the Applied Research Laboratories for controlling the sex of rats.

after repeated trials. While the offspring were all males, the number in the litters was below normal. The acid solutions gradually were decreased until experience showed 1 or 2 per cent to be the most satisfactory, although 0.5 per cent worked in two minimum strength matings. The soda solutions were reduced to either 2 or 3 per cent, with a minimum of 1 per cent effective in two cases.

An ordinary medicine dropper was selected as the most satisfactory instrument for douching.

The estrus cycle of each female was followed by daily examinations of vaginal smears. On the day at which estrus or "heat" was at a peak, the mother rat was douched. As much of the acid or alkaline solution as possible was introduced into her vagina and aspirated in and out at least three times to insure saturation of every fold and crevice in the vaginal tract.

Immediately after the douching the female was placed with a male in a small breeding cage located in a quiet darkened room. At the end of two hours the cage was examined for evidence of copulation as demonstrated by the presence of a vaginal plug. In the absence of a plug the douching process was repeated and the pair again placed together.

This procedure was repeated throughout the day at intervals of two hours until a vaginal plug was found. If no mating occurred, further attempts were postponed until another peak of estrus.

Selection and Distribution—Every livestock breeder knows that there are "chronic female-bearers" and "chronic male-bearers" among domesticated animals. Some parent animals consistently sire or bear a preponderance of offspring of the same sex in one mating after another and sometimes throughout their lives.

This tendency in mother rats was found by Dr. Helen D. King of Wistar Institute and confirmed by the litters born during *The News* feeding experiment. It is due apparently to the natural pH of the sperm-carrying fluid in males and the vaginal fluids in females. In consequence of this known tendency in female rats, the litter records and distribution of the Rat Experiment No. 1 females are significant. "Chronic female-bearers" were designated for alkali douches and the "chronic male-bearers" for acid douches, as follows:

For Acid Treatments (Previous litters)			*For Alkali Treatments* (Previous litters)		
Rat No.	M	F	Rat No.	M	F
A1673	11	5	8437	7	9
A845	8	7	A1615	5	6
A1532	9	7	A2416	6	9
A1229	14	4	A120	5	10
A5057	17	15	A1401	8	8
A1203	11	6	A9908	7	9
A3817	10	5	A2223	6	6
A2360	13	8	A5041	5	8
A2179	12	11	A4004	9	11
A2676	11	3			
A4648	9	6			
Totals	125	77	Totals	58	76

Thus the mother rats designated for acid treatments had previously borne a ratio of 162.3 males to 100 females, while those specified for alkali treatments had borne a ratio of 76.3 males to 100 females. In other words, the cards of nature were stacked against the success of the experiment. The attempt would be to make the "chronic female-bearers" bring forth male pups and the "chronic male-bearers" female pups.

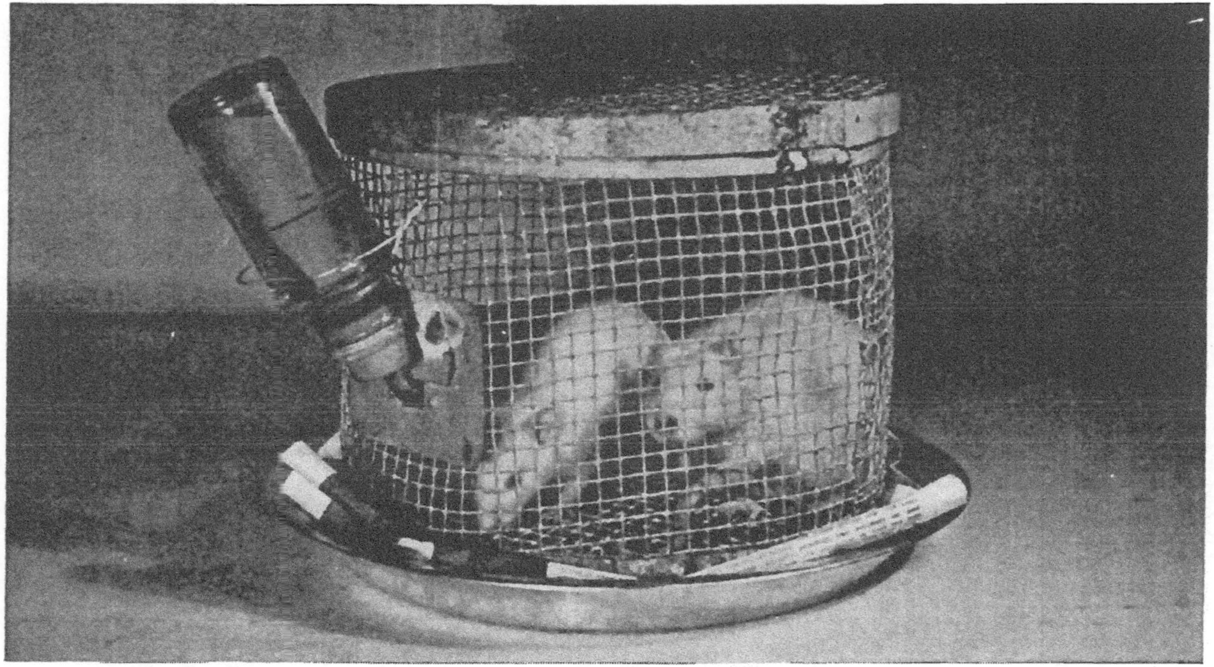

FIGURE 22.—RATS IN A MATING CAGE

Pair of animals used in Rat Experiment No. 1 at Dayton, N. J.

Results of Rat Experiment No. 1—*The News* up to this time had objectively reported the work being carried out at the Dayton laboratories and had focused public and scientific attention upon the experiment. Results were awaited with eager expectancy.

The first litter of baby rats arrived August 18, 1938, from mother rat No. 8437, a former "female-bearer," douched with a 5 per cent solution of soda. Her litter consisted of 3 males and 0 females. The following day soda-douched mother No. A1615 bore 4 males and 0 females. On September 9 the first acid-treated mother, No. A1673, gave birth to 5 females and 1 male.

And so it went to the end. Sex majorities in every one of the 20 litters were correctly predetermined by the Dayton scientists!

The sex ratio of pups in the acid-douched group was 24.6 males to 100 females and in the alkali-douched group 300.0 males to 100 females. Not only had ratios inexplicable under any natural circumstances been attained, but every single "female-bearer" had been artificially turned into a "male-bearer" and vice versa. The complete record follows:

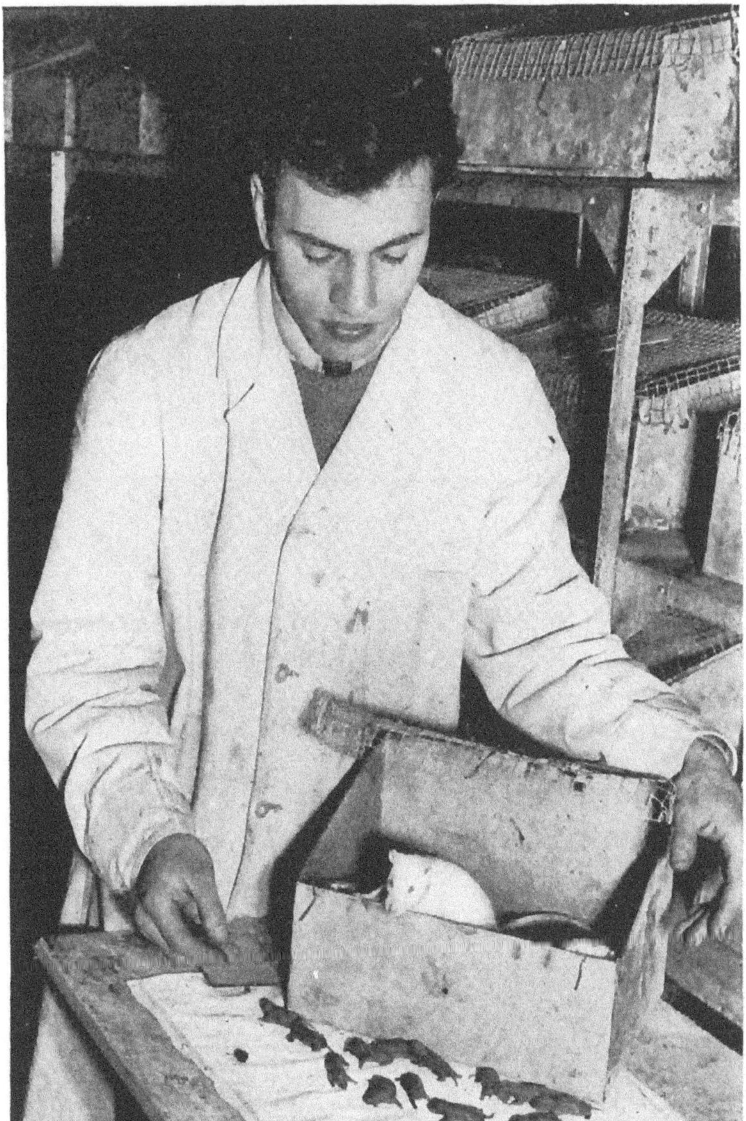

FIGURE 23.—CHECKING A RAT LITTER

The utmost care was taken in The News experiments to keep accurate records on sex-controlled animals.

Acid-Douched				*Alkali-Douched*			
Rat No.	Pct. Sol.	M	F	Rat No.	Pct. Sol.	M	F
A1673	1.0	1	5	8437	5.0	3	0
A845	0.5	2	6	A1615	5.0	4	0
A1532	1.0	1	6	A2416	2.0	9	1
A1229	0.5	3	5	A120	1.0	6	3
A5057	2.0	2	6	A1401	2.0	6	3
A1203	1.0	0	9	9908	1.0	6	5
A3817	2.0	0	5	A2223	2.0	4	0
A2360	2.0	2	6	A5041	2.0	6	2
A2179	2.0	1	5	A4004	3.0	4	2
A2676	2.0	3	7				
A4648	2.0	1	5				
Totals		16	65	Totals		48	16

The element of chance or luck had been eliminated and the last shred of doubt erased. Artificial sex control, in rats at least, was proved a success.

The Clinching Reversals—The Dayton experts now proposed and *The News* financed a laboratory venture induced by the query: Will a mother rat, once treated, respond to the opposite treatment?

Fifteen of the original 20 experimental females were available for re-breeding and were used for "reversal" re-tests. This time the treatments were reversed, that is, the acid-douched mothers were alkali-douched and the alkali-douched mothers were acid-douched. The cards again were stacked against success. The astoundingly accurate results follow:

Alkali-Douched (Previously acid-douched)				*Acid-Douched* (Previously alkali-douched)			
Rat No.	Pct. Sol.	M	F	Rat No.	Pct. Sol.	M	F
A845	2.0	4	2	A2416	2.0	1	3
A1532	1.0	7	3	A120	1.0	0	7
A1229	2.0	6	0	A1401	2.0	1	8
A5057	2.0	8	0	A9908	1.0	4	6
A1203	1.0	4	3	A5041	2.0	1	5
A3817	2.0	4	2	A4004	3.0	0	6
A2179	2.0	5	3				
A2676	2.0	7	3				
A4648	2.0	4	1				
Totals		49	17	Totals		7	35

Let us now lump the litter records of the same 15 females used in the experiment and observe the violent changes after treatments were reversed:

Sex Ratio
(Same 9 Females)

Acid-douched	24.1 males to 100 females
Alkali-douched	288.2 " " 100 "

(Same 6 Females)

Alkali-douched	231.3 males to 100 females
Acid-douched	20.0 " " 100 "

It was interesting to note that in the reversals an even greater proportion of the desired sex was obtained than in the original experiment. This probably was due to a regression toward the normal tendencies of the females.

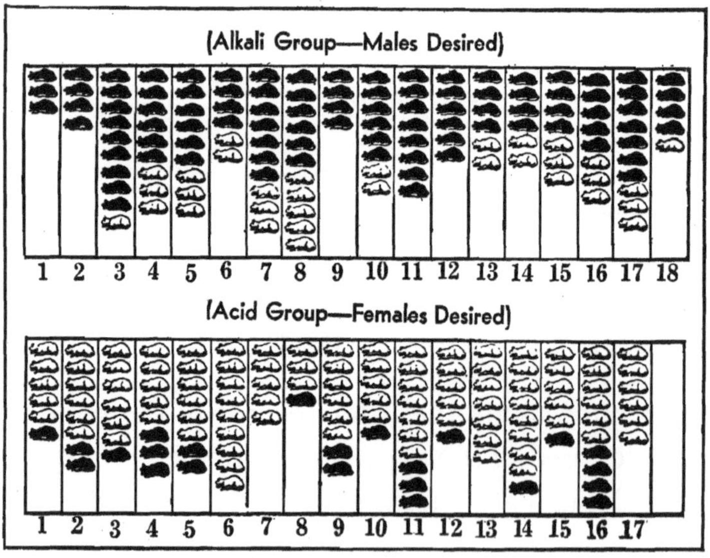

FIGURE 24.—RESULTS OF RAT EXPERIMENT NO. 1

All litters produced in this series of tests had desired sex preponderances. Black symbols represent males and white symbols females.

The reversals made it possible for *The News* to announce that not only could sex majorities be arranged at will in advance of conception, but that they could be changed in subsequent litters by

FIGURE 25.—TWO "REVERSAL" MOTHERS

Each of these animals was treated with both acid and alkali. All four litters had the desired sex majorities.

reversing the treatment. This would mean that a woman may have children of either or both sexes in the order she wishes if the method becomes wholly practical for human use.

The News Research Extended—All America was stirred as the Dayton experts reported success after success. When *The News* made the reports public many other newspapers and periodicals reprinted and commented upon them.

A war of words raged across the continent and *The News* in a single issue printed the comments of 200 physicians, livestock geneticists, sociologists and others. Livestock men almost unanimously acclaimed the achievement and a torrent of demands by breeders for specific data followed. Scientists generally urged more extended research.

The News laid the matter before officials of the United States Department of Agriculture who suggested the use of more rats and a trial with another species of animal–guinea pigs or rabbits. Thereupon *The News* launched a wholly new and widened series of tests known as Rat Experiment No. 2 and Rabbit Experiment. These began simultaneously in November, 1938.

New Rat and Rabbit Experiments—A total of 100 female rats was treated in this series of confirmatory tests which were completed in July, 1939. The method of treatment was identical with that followed in Rat Experiment No. 1. Fifty of the females were douched with acid and 50 with alkali.

This time, however, the 100 females were selected and divided at random with no reference to their breeding histories. One hundred other females were designated as controls for purposes

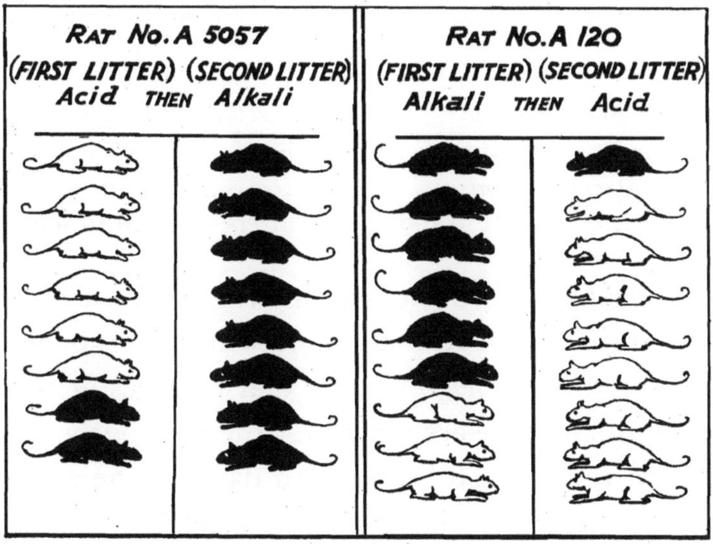

FIGURE 26.—TWO CLINCHING REVERSALS

Results were uncannily accurate when treatments of the same female rats were changed from acid to alkali and vice versa.

of comparison. Controls in such an experiment are animals fed and handled exactly like the test animals but allowed to breed normally without treatment.

The results, while not 100 per cent correct,

supplied statistical evidence more conclusive than the earlier rat experiment due to the larger number of animals involved. Out of 100 litters produced there were 94 with predetermined majorities, 5 ties and only 1 failure. The sex ratios follow:

Acid-douched	*Alkali-douched*
50 litters. . 35.2 M to 100 F	50 litters 341.6 M to 100 F
Untreated Controls	*Untreated Controls*
50 litters. . 96.0 M to 100 F	50 litters 100.0 M to 100 F

In addition to demonstrating the rat work on a bigger scale, Durfee also knocked out a belief advanced by some reviewers that the ages of the dams might have accounted for the previously obtained results. He distributed dams of the same age through the alkali, acid and control groups and tabulated their litters separately. There was no significant deviation among them from the group as a whole.

Durfee said, "The effect of the acid and alkali is the controlling factor even though the age of the dam may exert some slight influence on the results."

The exact procedure and detailed results in

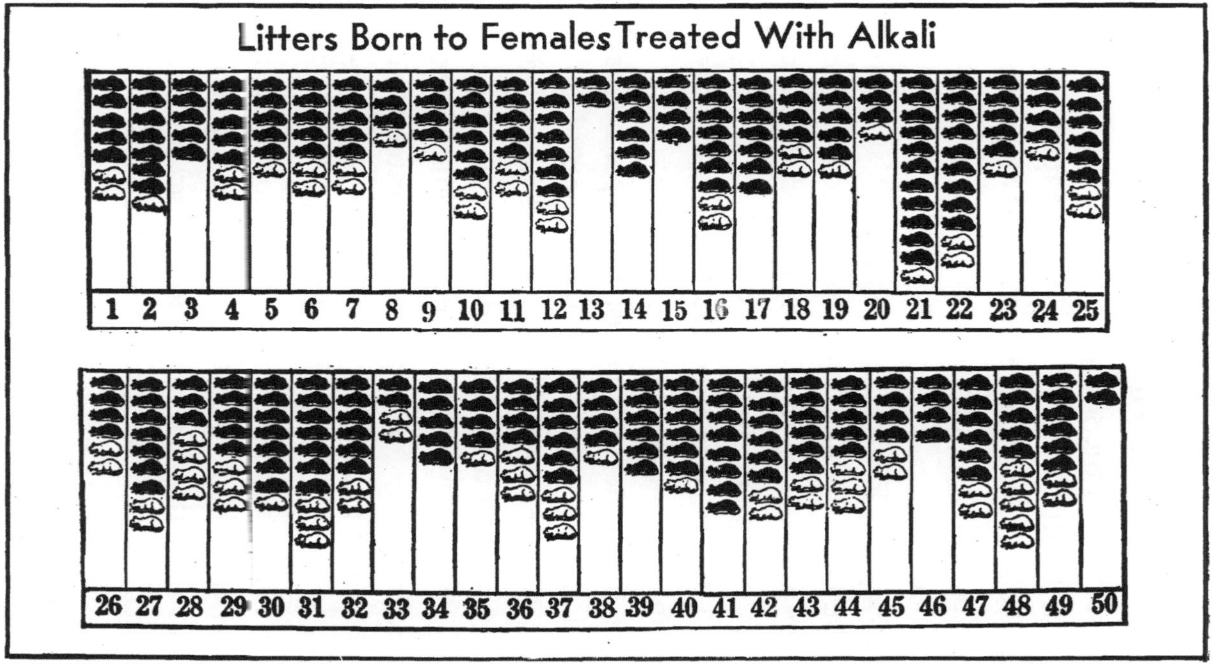

FIGURE 27.—FIRST HALF OF RAT EXPERIMENT NO. 2

Each of these litters was produced by a female douched with a soda solution. Black figures represent males and white figures represent females.

the rabbit experiment, which opened the door for breeders to control the sex of an animal raised for profit, will be discussed in the next chapter. It will only be said here that out of a total of 40 rabbit matings nearly 80 per cent of the desired sex was obtained.

Proof Beyond Reasonable Doubt—A summation of the experimental work sponsored by *The News* over a period of 18 months showed that sex majorities in a grand total of 167 out of 175 rat and rabbit litters were predetermined with only 7 male-female ties and 1 failure.

For the first time in America the experiments confirmed the discovery of Unterberger and upheld the achievements of the Taylors. They piled up a staggering volume of evidence proving beyond any reasonable doubt that the riddle of sex control was solved, for two kinds of animals at least, and paved the way in America for predetermination in all mammals, including man.

Durfee's final report to *The News* concluded: "The essential scientific question is answered. We have controlled the determination of sex in rats and rabbits. The tasks ahead are those of perfection and adaptation only."

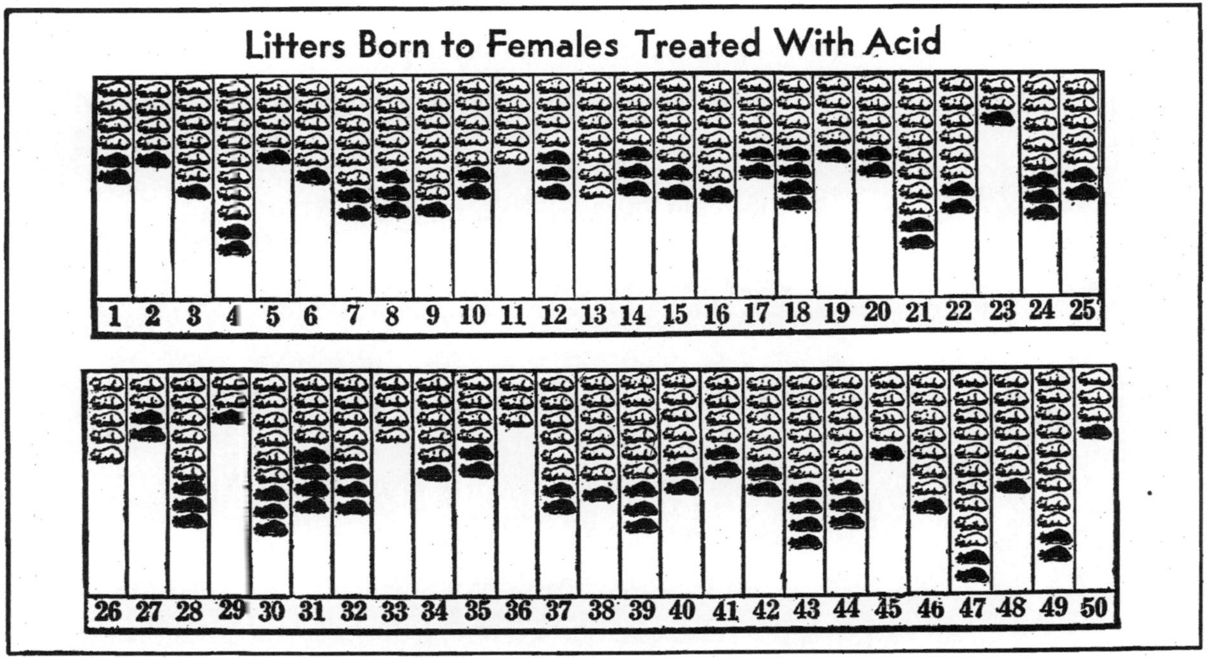

FIGURE 28.—SECOND HALF OF RAT EXPERIMENT NO. 2

Each of these litters was produced by a female douched with a lactic acid solution. Black figures represent males and white figures represent females.

FIGURE 29.—TYPES OF EXPERIMENTAL RABBITS

Four animals used at Dayton, N. J., in America's first successful sex control experiments with rabbits.

VII

METHODS FOR SMALL ANIMALS

Sex Differences in Animals—The acid-alkali method has been applied successfully to predetermine sex majorities in rats, rabbits, dogs, cattle, horses and human beings. It should work with equal certainty on cats, mink, foxes, goats, swine and all other mammals. In its present stage it definitely will not work on birds, such as chickens, geese, ducks and turkeys, nor on fish or amphibians, such as frogs.

The reasons for these differences are to be found in the distribution of the sex-determining chromosomes. In mammals the sperms determine sex. But in birds, most fish and some insects the ovum carries both the X and Y sex-determining chromosomes.

Since alkali and acid affect only the fertilizing power of sperms it is obvious why they work in mammals but not in poultry, fish and frogs.

Materials for Rabbits—By far the most research work has been done on rats, especially in the two experiments sponsored by *The News*. Equally good results, however, were attained at the Applied Research Laboratories with rabbits which, because of their commercial value, are of more interest to practical breeders.

Does recently have been in demand by laboratories for human pregnancy and hormone tests. Females also are usually preferred in larger numbers for breeding. Males, because of their greater weight, are desired for meat production.

In his laboratory Durfee developed equipment for rabbits similar to that used for rats, but with certain modifications due to the anatomical differences in these animals. The douching device was a hard rubber uterine syringe. It had a capacity of 10 cc with a slender nozzle about 8 cm long, about 5 mm in diameter, and with a 1 mm bore. The end of the nozzle was curved slightly 3 cm from the tip to facilitate entrance. For the same reason, the end was given a thin coating of vaseline before it was used.

Douching the animals was a simple matter. The rabbit was secured to a common animal holder. The syringe was carefully inserted into the vagina.

FIGURE 30.—SEX-CONTROLLED RABBIT LITTER

One of the first females to be successfully treated and her young. All were healthy and normal in every way.

If the animal was in heat, the syringe could be inserted nearly to the uterus with no extreme pressure being used. The amount of solution used varied with the size of the animal, usually from 4 to 8 cc being sufficient for a treatment.

Preliminary matings showed that a 2 per cent solution of soda was the nearly correct concentration. Alkaline solutions of 3 per cent and higher reduced the litter size considerably, while a 1 per cent failed to produce the desired preponderance of males. In the same manner, a 2 per cent solution of lactic acid was found the most suitable to obtain female offspring.

These solutions can be mixed by any pharmacist at a trifling cost. They are made up by weight from C.P. or U.S.P. sodium carbonate and lactic acid (85 to 90 per cent lactic acid is the most commonly available material) and distilled water. It was found advisable to prepare fresh solutions weekly since they tend to grow moldy and also gradually change in reaction if kept longer than this period. The solutions should be kept in amber colored bottles if held even for a few days.

Treatment of Rabbits—Estrus in female rabbits does not occur at regular intervals as in rats

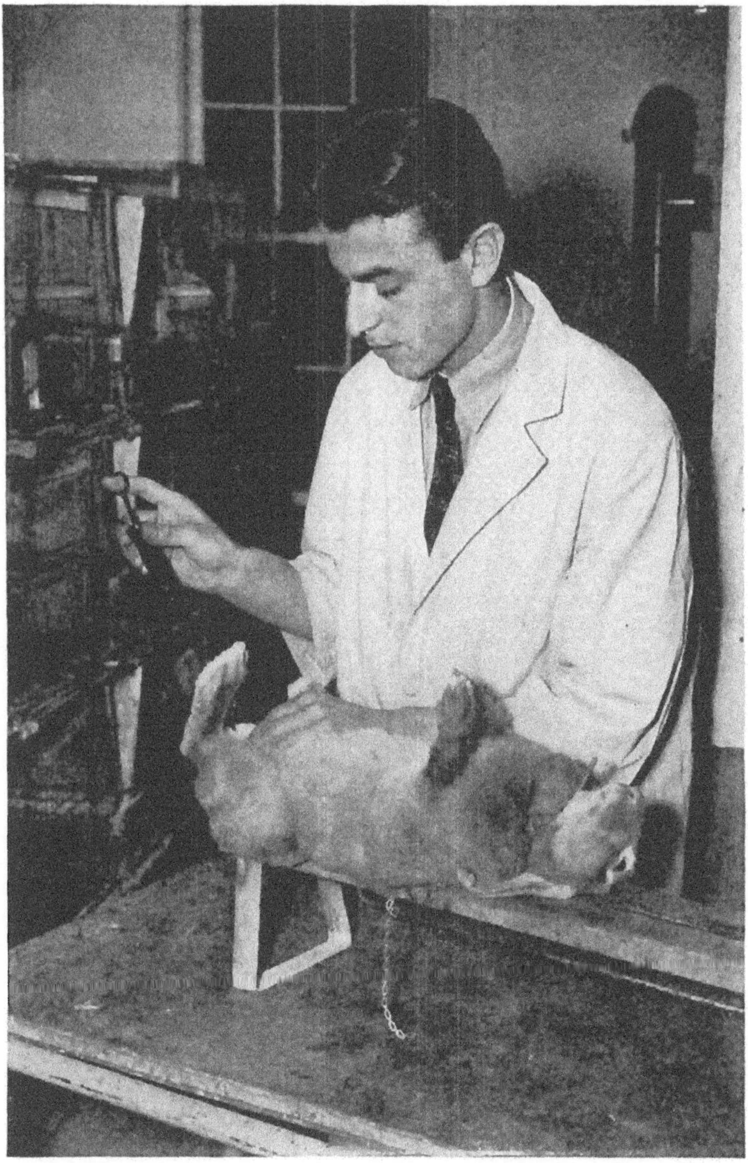

FIGURE 31.—DOUCHING A RABBIT

A biologist at the Applied Research Laboratories demonstrates the use of a syringe.

and most other animals. It was necessary, therefore, to bring about estrus a few days before the actual matings were to be undertaken by placing the does in hutches adjacent to a buck.

Just before mating the female was douched with the appropriate solution prepared in the manner previously described. As much of the solution as possible was introduced into the vagina and aspirated in and out at least three times. Immediately thereafter the doe was placed in the buck's cage and kept under observation to be sure that coitus occurred.

If no mating was observed within one hour, the douching of the doe was repeated. In each case

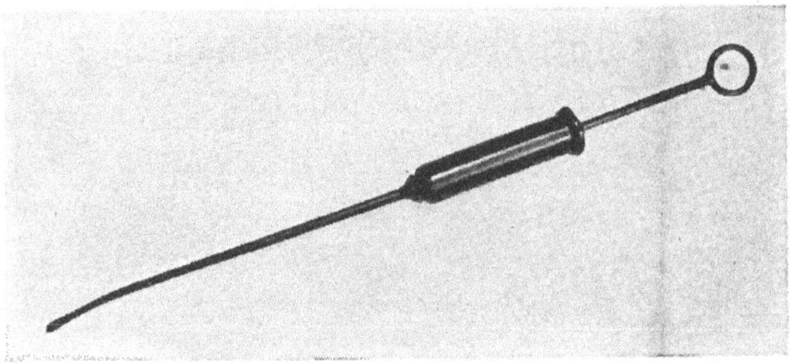

FIGURE 32.—SYRINGE FOR RABBITS

The instrument for douching rabbits used at the Applied Research Laboratories.

the buck was allowed to serve the doe at least two or three times if this was accomplished within the maximum time limit of one hour.

Results of Rabbit Experiment—Since about the same outcome can be expected by the average breeder, the complete results of *The News* rabbit experiment are set forth here:

Alkali-douched				*Acid-douched*			
Rabbit No.	Pct. Sol.	M	F	Rabbit No.	Pct. Sol.	M	F
709	1	1	0	704	1	2	6
740	1	6	3	733	1	3	4
712	1	4	4	702	2	2	5
706	2	5	0	714	2	1	4
701	2	3	0	713	2	1	4
715	2	8	0	723	2	2	4
717	2	8	1	735	2	2	10
731	2	9	1	732	2	3	7
726	2	6	3	745	2	3	4
760	2	2	2	731	2	2	5
741	2	6	1	754	2	2	4
721	2	5	2	748	2	1	4
744	2	7	1	736	2	0	3
757	2	6	2	750	2	0	1
762	2	6	0	755	2	2	8
761	2	2	1	758	2	0	5
737	2	3	0	759	2	1	4
747	2	8	2	710	3	1	7
746	2	3	1	743	3	0	2
703	3	2	0	724	3	0	2
Total		100	24	Totals		28	93

All litters, a total of 40, were controlled except 2 in which there were male-female ties. There was 1 reversal. Doe No. 731, douched with soda, bore 9 males to 1 female. Later, douched with acid, she bore 5 females to 2 males.

In the alkali group, the ratio of males to females was 416.6 to 100 and in the acid group 30.1 to 100. Analyzed another way, the per cent of the desired sex obtained was 80.6 from alkali and 76.9 from acid, an average of 78.8 for all the rabbits treated. It may be concluded that in its present stage the acid-alkali method will give a rabbit breeder more than three-fourths of the sex he desires.

The Touchette Dog Method—The first sex-controlled dog litters definitely recorded in the United States belonged to Mrs. Mary Touchette of New York City. She obtained desired sex majorities by douching two Springer Spaniel bitches, following basically the method reported by *The News* scientists.

Mrs. Touchette procured from a pharmacist a quart of 2 per cent soda solution in distilled water and a quart of 2 per cent lactic acid solution. She also bought an ordinary hard-rubber gooseneck syringe of the type used by women. To the nozzle

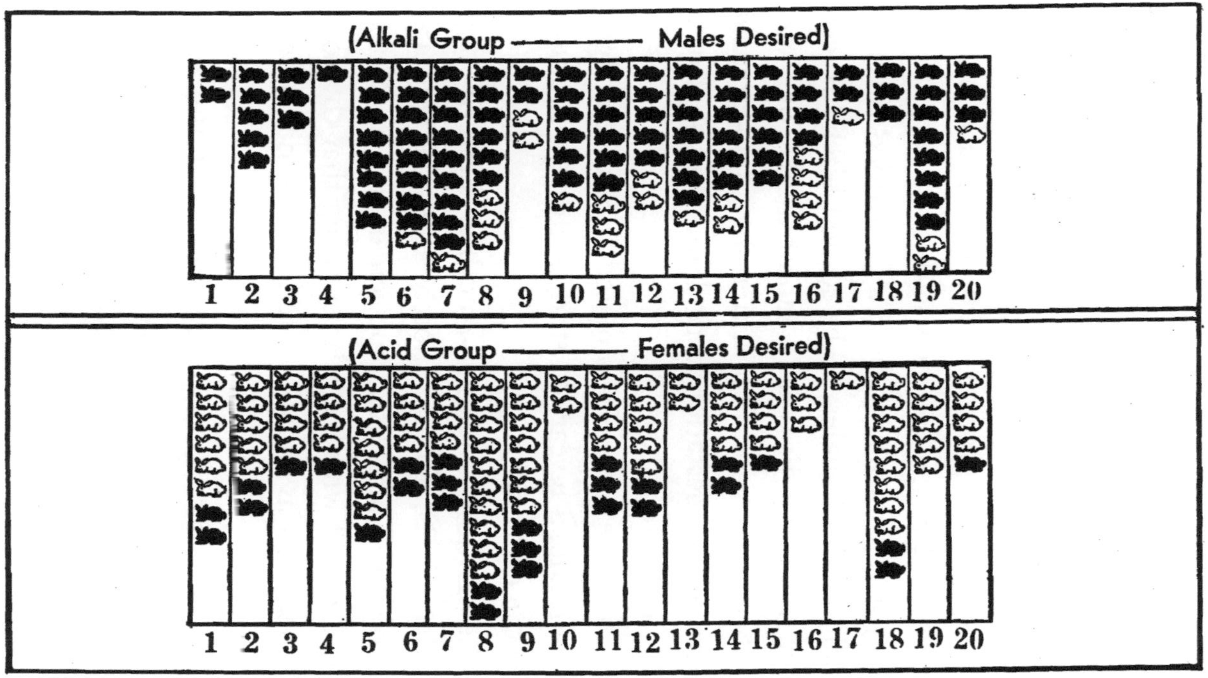

FIGURE 33.—RESULTS OF RABBIT EXPERIMENT

All litters, a total of 40, were sex-controlled except 2 in which there were male-female ties.

she applied a thin coating of vaseline. She warmed the solutions to body heat and douched the bitches in a bathtub.

The first, Blossom Lynn, was 3 years old and had borne one previous litter of 6 males and 3 females. She was douched with soda and after a 63-day gestation period bore a controlled litter of 4 males and 3 females. The second bitch, Skeetsbarry of Merry T, was 15 months old. She was douched with acid. The litter—her first—arrived after a 61-day gestation period and consisted of 2 males and 7 females.

Each of the bitches was mated to Prince of Igor of Avondale, a 9-year-old stud, within 45 minutes after the douchings.

The Dillon Litters—Another Eastern dog breeder, T. D. Dillon, shortly after the publication of the Touchette experience in *The News,* came forward with records revealing that the control method was equally effective for other breeds of dogs.

Like Unterberger, Dillon had used the soda douche years before to make females fertile. He saw the significance in *The News* tests and resolved

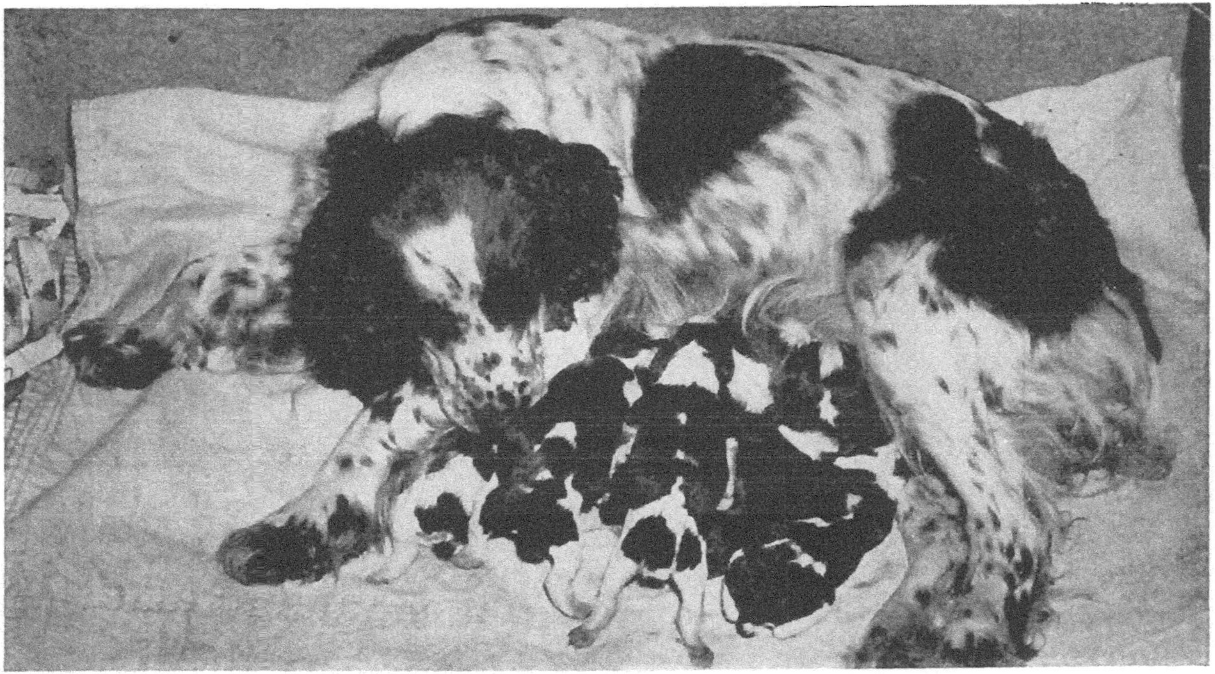

FIGURE 34.—AMERICA'S FIRST SEX-CONTROLLED DOGS

Blossom Lynn, owned by Mrs. Mary Touchette, was treated with soda and whelped a majority of male pups.

to try sex control. He too used a common hygiene syringe and 2 per cent solutions.

He treated first with soda an Airedale with the

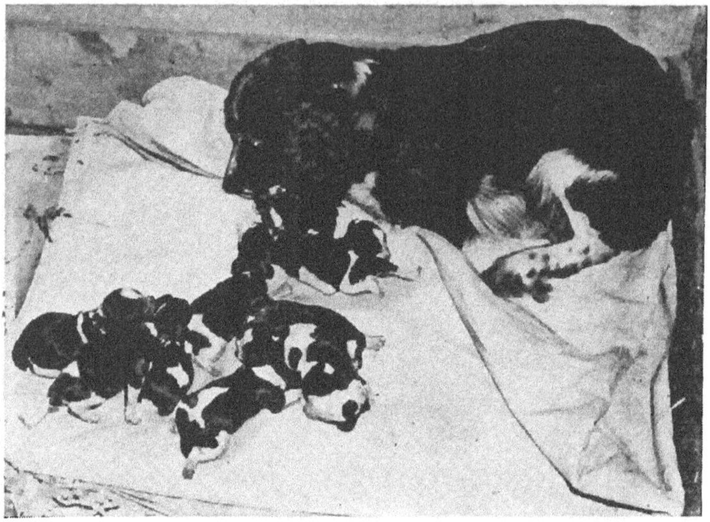

FIGURE 35.—ANOTHER CONTROLLED LITTER

Skeetsbarry with her 2 male and 7 female pups. She was treated with a solution of lactic acid.

kennel name of Babs, a chronic "female-bearer." Her previous litter had consisted of 6 females and 2 males, and the one before that was also heavily female. Her controlled litter was comprised of 7 males and 6 females. In rapid succession Dillon treated other bitches as they came in heat with these results:

Acid-douched			*Alkali-douched*		
Breed	M	F	Breed	M	F
Springer Spaniel...	2	5	Airedale	7	5
Police	0	5	Police	4	1
			Irish Terrier	5	2
			Irish Terrier	5	0

Dillon, like Mrs. Touchette, reported that the bitches behaved well during the treatment and remained normal in every way thereafter. All the

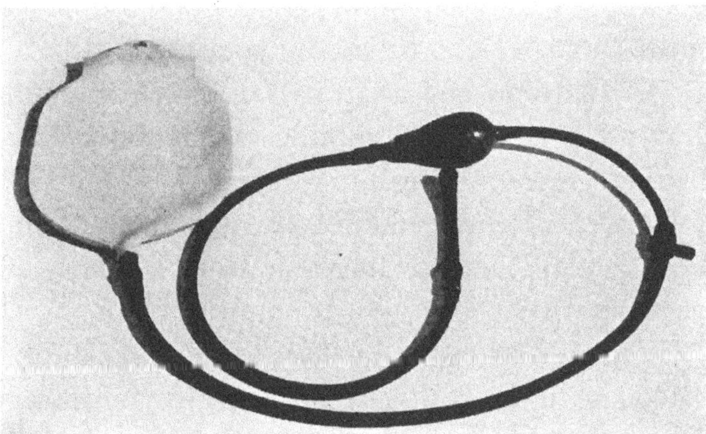

FIGURE 36.—SYRINGE FOR DOGS

Douching apparatus for dogs recommended by the Applied Research Laboratories.

pups obtained by both breeders were healthy at birth and remained so.

More Dogs Controlled—The Touchette and Dillon work drew hundreds of inquiries and scores of other dog breeders tried the method. Meanwhile, George E. Bayer at the Mountainside Great Dane Kennels, Mount Carmel, Pa., joined in the pioneering efforts.

During 1936 he had attempted with considerable success to influence sex by underfeeding the stud and bringing the bitch to prime physical condition on the theory that the sexually stronger female, due to the alkalinity of cervical fluid given forth during coitus, is likely to conceive males.

He then douched a Great Dane bitch, Blondl, with a 5 per cent soda solution 30 minutes before mating. She gave birth to a litter of 4 males and 2 females. Considering the strength of the solution, it seemed surprising that so many puppies were obtained.

William Godfrey of Yonkers, N. Y., next duplicated one of the Dillon experiments by transforming a "female-bearer" into a "male-bearer." His thoroughbred Irish Terrier bitch, Kilderry Lady, had borne three litters before treatment, the first

FIGURE 37.—FIRST SEX-CONTROLLED POLICE DOG

T. D. Dillon, a pioneer in sex control work, showing predetermined litter obtained after treating his bitch, Tara, with soda.

10 to 1 female, the second 7 to 1 female, the third 6 to 0 female—all from one stud, Padraig Roo of Armagh. He douched her with 2 per cent soda and using the same stud procured a fourth litter with a 7 to 3 male preponderance.

Fresh reports from dog breeders came in rapidly after the sequence of 10 controlled litters. These reports piled evidence upon evidence that predetermination for dogs, as well as rats and rabbits, had been accomplished.

Technique for Dogs—After a study of the equipment and procedure used by the various breeders, Durfee concluded that the syringe best adapted to dogs is a common hygiene syringe with a water reservoir bag attached, obtainable in any pharmacy.

Before douching, he recommended that the end of the syringe should be covered with vaseline and the solutions should be warmed to body temperature. Care should be taken to raise the reservoir bag high enough to insure enough pressure for the douche.

The treatment can be considered completed when the solution begins to overflow from the vaginal tract. Here again if mating does not

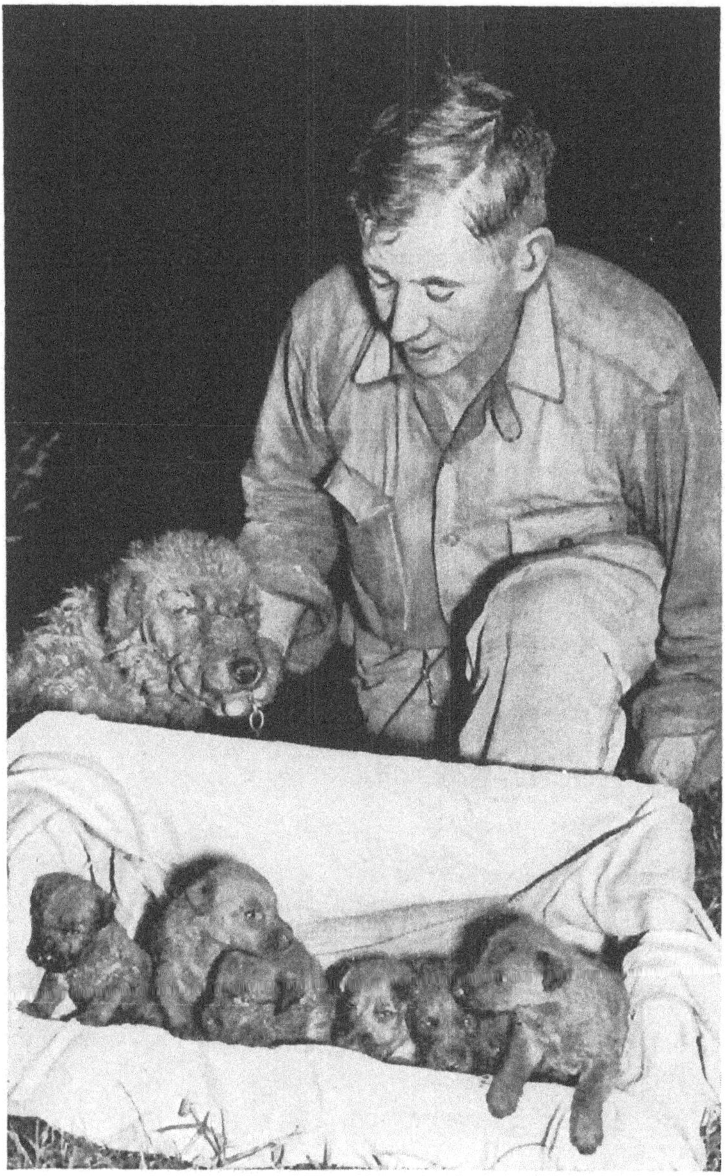

FIGURE 38.—IRISH TERRIER TREATED SUCCESSFULLY
T. D. Dillon inspecting 7 pups born to an Irish Terrier, Peggy.

occur within an hour, a second douche is advisable.

Other Small Animals—Breeders of cats, guinea pigs, mink and other small animals are taking interest in the sex control experiments but in smaller numbers than rabbit, dog, cattle and horse raisers.

While actual results one way or the other still are lacking, the repeated confirmations with rabbits and dogs leads inevitably to the conclusion that alkali-acid control eventually will be adapted to every mammal bred under human supervision.

VIII

METHODS FOR LARGE ANIMALS

The Problem of Adaptation—The sex control of cattle and horses today is in the practical stage but statistical data, while specific and clearly indicative, are less abundant than for smaller animals for reasons apparent to any breeder.

Rats and rabbits cost little and breed rapidly. Their gestation period lasts only 3 weeks. Dogs, too, are comparatively inexpensive and bear litters in slightly more than 2 months. Cattle and horses, on the other hand, represent heavier investments and generally bear single offspring at relatively longer intervals. The gestation period for the cow is a little more than 9 months, and for the mare about 11 months.

From another viewpoint, adaptation to larger animals is simplified by the wealth of detailed knowledge about sex determination in human beings which, like cattle and horses, fall in the large animal category and usually produce one offspring after a gestation period of 9 months.

However, there exists no fundamental physiological difference in the natural determination of sex in large and small mammals as the sperm is the sole governing factor in each and hence the acid-alkali theory applies to them all. The differences concern technique rather than basic principles.

Soda and Sterility—The earliest recorded observation of chemical influence upon the sex of cattle, as we know, was that of Unterberger in the 1920's. He noted that cows treated with soda vaginal washings for sterility usually bore bull calves. This led him to his unprecedented experiments with women.

Strangely enough, other cattle breeders recently have come forward with stories of bull calves born to soda-rinsed cows. Gilbert Lanterman of Lyons, N. J., upon reading of *The News* experiments, revealed that he had douched his barren cows for three years and that 12 bulls out of 14 calves resulted.

So far as is now known, however, only Unterberger followed up the discovery. Perhaps one

FIGURE 39.—JERSEYS IN CATTLE EXPERIMENTS

These pure-blooded virgin heifers were treated at the Shelby County Penal Farm near Memphis, Tenn.

reason for this failure of others to further investigate the phenomenon has been the desire of dairymen, intent on increasing butter-fat production, for heifer calves. At birth a heifer in a dairy herd is worth twice as much as a bull and the average dairyman would, if he could, produce them in a ratio of 10 to 1. Although cattle ranchers prefer males for beef steers, their interest in scientific breeding has been no less intensive than that of the milk cow men.

New Cattle and Goat Experiments—During 1939 when the center of progress with the acid-alkali control method was shifted to America by *The News* success with rats and rabbits, cattle and goat raisers awakened to its significance.

Experiments started in all parts of the country, including one at the Shelby County Penal Farm, Memphis, Tenn., under the direction of Gale Carr, former superintendent of the Dairy and Poultry departments. Carr treated a dozen pure blooded virgin Jerseys. His first controlled calf, a heifer, was born to an acid-douched cow in August, 1939. Soon thereafter he obtained a bull calf from an alkali-douched cow.

Goat breeders deluged *The News* with inquiries

and several started treatments from which important results were expected.

First Success With Horses—How the old-fashioned soda remedy for sterility beckoned

FIGURE 40.—BULL USED IN BREEDING TEST COWS

Bouncing Bet's Observer, a prize animal, which sired calves in tests at the Shelby County Penal Farm in Tennessee.

Unterberger and others toward the secret of sex control already has been repeatedly mentioned. The same sequence occurred in the work of Dr. J. D. Cummings, a veterinarian and horse breeder of Alberton, Prince Edward Island, Canada.

One evening in May, 1937, a neighbor brought to Doctor Cummings an apparently sterile mare in which the veterinarian found a catarrhal condition of the cervix uteri, seemingly the cause of her barrenness. Soda was well known to him as a remedy. He placed 380 grains, apothecary weight, into a capsule and inserted it well into the neck of the womb. The next morning, 12 hours later, a stallion served the mare Pansy Aubrey. Eleven months afterward she gave birth to a male colt.

"As soon as I knew this mare was with foal, I began to use the treatment indiscriminately with no idea of its effect on sex determination, but merely in an effort to assure fertility," Doctor Cummings explained. "The next spring, however, it suddenly struck me one day that out of the 20 mares so douched, 15 of them had male colts! Right then I knew I had something."

Doctor Cummings eagerly searched his ledgers and tabulated the dates of treatment service and foaling.

He had obtained 15 males from 20 mares treated during 1937. Six foals from seven mares treated later were males. Thus, from a total of 27 treated mares, Doctor Cummings obtained 21 males and

6 females, an abnormal sex ratio attributable only to the soda. The 1937 record follows:

Mare	*Treatment Date* (1937)	*Sex of Colt*
Pansy Aubrey	May 3	Male
Nellie	May 11	Male
Winnie	May 12	Male
Unnamed	May 18	Female
Unnamed	May 31	Female
Peggy	June 9	Male
Betsy J.	June 9	Male
Unnamed	June 12	Male
Unnamed	June 16	Male
Helen	June 21	Female
Unnamed	June 30	Male
Lady Forbes	July 6	Male
Winona	July 6	Male
Unnamed	July 10	Male
Maggie	July 12	Female
Unnamed	July 13	Male
Unnamed	July 16	Female
Unnamed	July 19	Male
Unnamed	July 22	Male
Unnamed	July 28	Male

Soon after learning of *The News* experiments, Doctor Cummings started tests with lactic acid. Other horse breeders meanwhile began work with solutions of both the male and female-begetting chemicals applied with douches instead of capsules.

An Early Swine Test—A limited experiment with swine was made in 1932 in the Department of Genetics, University of Wisconsin, in cooperation with the Animal Husbandry Department of the Wisconsin Agricultural Experiment Station.

Thirty-two sows were treated, the vagina of each being flushed before breeding with 50 cc of a solution of 0.5 per cent soda. Thirty-six other sows in the same herd and bred during the same period served as controls. Of the treated sows 19 conceived after the first mating and bore 170 pigs of which 83 were male and 87 female. The controls bore 36 litters of 141 males and 148 females.

The results of this test, although meager, discouraged experimental work in the United States during the post-Unterberger era. The soda apparently was not effective.

In the light of subsequent experiments the explanation for the failure seems clear. The 0.5 per cent solution was entirely too weak. With both rats and rabbits Durfee found it necesary to use 2 and 3 per cent solutions, containing 4 to 6 times

FIGURE 41.—ALKALI GAVE HER A MALE COLT

Trixie Volo, a famous racing mare, trained and treated with soda capsule by Dr. J. D. Cummings of Alberton, P.E.I., Canada.

the 0.5 per cent used in the Wisconsin test on pigs.

It was also pointed out that the semen may not have been reached by the solutions due to ordinarily deep penetration of the male organ during the mating of swine.

These differences alone make new swine experiments desirable and there is every reason to believe that they will be successful as with other mammals.

Technique for Cows, Horses and Goats—Application of the acid-alkali method to the large animals—cows, horses, goats—as recommended by the Applied Research Laboratories is a fairly simple procedure.

A hard rubber syringe about 1.5 inches in diameter and 12 inches long is used. Several holes along the inner 6 inches of the syringe allow the solutions to be ejected over a wide surface. A rubber tube attached to a reservoir bottle is joined to the syringe. A clamp on the tube is used to regulate the amount of solution withdrawn. The reservoir bottle should hold about 2 to 5 gallons of solution. The reservoir should be placed high enough to insure a gentle but firm water pressure.

In the actual douching, it is important that the vagina be distended so that the solutions will reach every point in the vaginal lining. This is accomplished in one of two ways. A shield, about 5 inches in diameter, is placed on the rear end of the syringe. The syringe is then inserted into the vagina until the shield presses against the vulva. The introduction of the solution will then cause the vagina to distend slightly. Too much pressure should be carefully avoided.

The vagina may also be distended in the douching process by simply holding the lips of the vulva together when introducing the solution. In either case, the liquid is flushed out and a fresh quantity introduced. After flushing a second time, a final amount is introduced and again flushed out. The cow is then ready to be served. Here again if mating does not occur within an hour a second douche should be given.

The strengths of solutions used are lower than in the smaller animals. From the work that has been done, it has been indicated that a 0.5 to 1.0 per cent of acid and a 1.5 to 2.0 per cent solution of soda are the best concentrations.

Artificial Insemination—The rapid progress

made during the last few years with artificial insemination—impregnation of the female with male semen injected with a syringe—may soon play an important part in sex control.

Artificial insemination was first worked on dogs in the late 1700's by Lazaro Spalanzani, an Italian scientist. Now it is being used especially in cattle-breeding throughout the world. The method multiplies many times the usefulness of fine sires, puts cows in barn space formerly occupied by inferior bulls and eliminates the transmission of disease. Spermatozoa can easily be kept alive for 24 hours and have been preserved in an active state for much longer periods.

Artificial insemination raises the question: Why not treat the semen, rather than the vaginal tract, with soda or lactic acid? There is no theoretical reason why this cannot be done. The time may come when the breeder literally may be ably to obtain in a single container semen that will insure both the desired hereditary factors from the male and the desired sex in the offspring.

IX

BOYS OR GIRLS—WHICH?

The Human Issues—What will sex control do to the human race? Will it unbalance the sexes with ensuing disaster or will it regulate nature to the improvement of family life and public welfare? These are at once the most fascinating and controversial questions raised by the rapid advancement of the new science.

A good deal of dust befogs the subject—much of it kicked up by alarmists akin to those who viewed the telescope, chloroform, the surgeon's knife, the steam engine, vaccination and the airplane as interference with matters which were none of mankind's business.

Everyone agrees, of course, that sex control of livestock is of incalculable benefit to farmers. But the mind inevitably jumps ahead to the time when parents can deliberately choose between a boy and a girl prior to each conception and when the mode of procedure will be commonly known.

While that time has not yet arrived, it is swiftly drawing near.

Snap judgments and curbstone conclusions of fear-mongers will not check the never ending struggle toward the truth. The basic facts about sex control have been found and they will prevail. However, it may be well to examine here the most popular views on the subject in the cold light of social history and factual evidence. It will be seen that many of them are nonsensical.

The Three Sex Ratios—The obvious presupposition of sex distribution is that human males and females universally are equal in number. That is a false presumption. Inequalities exist now.

There are three kinds of sex ratios—the primary sex ratio at conception; the secondary sex ratio at birth; and the tertiary sex ratio at maturity. It is the primary sex ratio which can be altered by the acid-alkali method, but natural circumstances already upset the secondary and tertiary ratios.

We have noted the fact that rats normally give birth to 105-6 males to 100 females and rabbits 104.6 males to 100 females. The male-female ratio at birth varies with species. In dogs it is

118.5, in swine 111.8, in cattle 107.3, in horses 98.3 and in sheep 97.7.

Among human beings there is a general excess of boys at birth, ranging from 103 to 107. But the preponderance is not evenly spread. The black races, for example, produce fewer males than whites. Masculinity increases in wartime. Urbanization favors femininity. The tertiary sex ratio also varies. Men exhibit an inherently inferior resistance to the stresses of the act of living and die at earlier ages than women.

These puzzling variations in human sex ratios have been the subject of intensive research and scientific explanations are rapidly coming forth from these studies.

The point stressed here is that the much-talked-about ideal of 50-50 sex distribution has never existed and does not exist now. Largely uncontrolled factors arbitrarily bring inequalities in crazy-quilt haphazard fashion. Artificial predetermination, even if it alters sex ratios, will merely do under intelligent supervision what unintelligent forces are doing now.

Monkeying With "Nature"—Often repeated by non-scientific people are the remarks: "Nature knows best" and "It is better to leave things as providence planned them."

A moment's reflection will show that many factors besides nature and providence affect the sex ratios, including illegitimacy, pre-natal neglect, infant mortality, war, and so on. Shall we refuse to do anything about these evils because thereby we would be tinkering with "nature"? Of course not. The fact is that neither nature nor providence can be held responsible for them. They are man-made and should be man-changed.

But let us assume that nature does, on the whole, "even things up" by providing an excess of males at birth and killing off the excess later in life. Does it follow that an even balance of the sexes should prevail in all groups all the time?

In democratic America a superficially-minded person is prone to predict that war-like dictators would raise huge armies of men for cannon-fodder. Should such a foolish and suicidal end be desired by bloodthirsty dictators, they could achieve it now by the simple expedient of slaughtering off females at birth.

Other people foresee nations of Amazons with men as pariahs tolerated only for procreative pur-

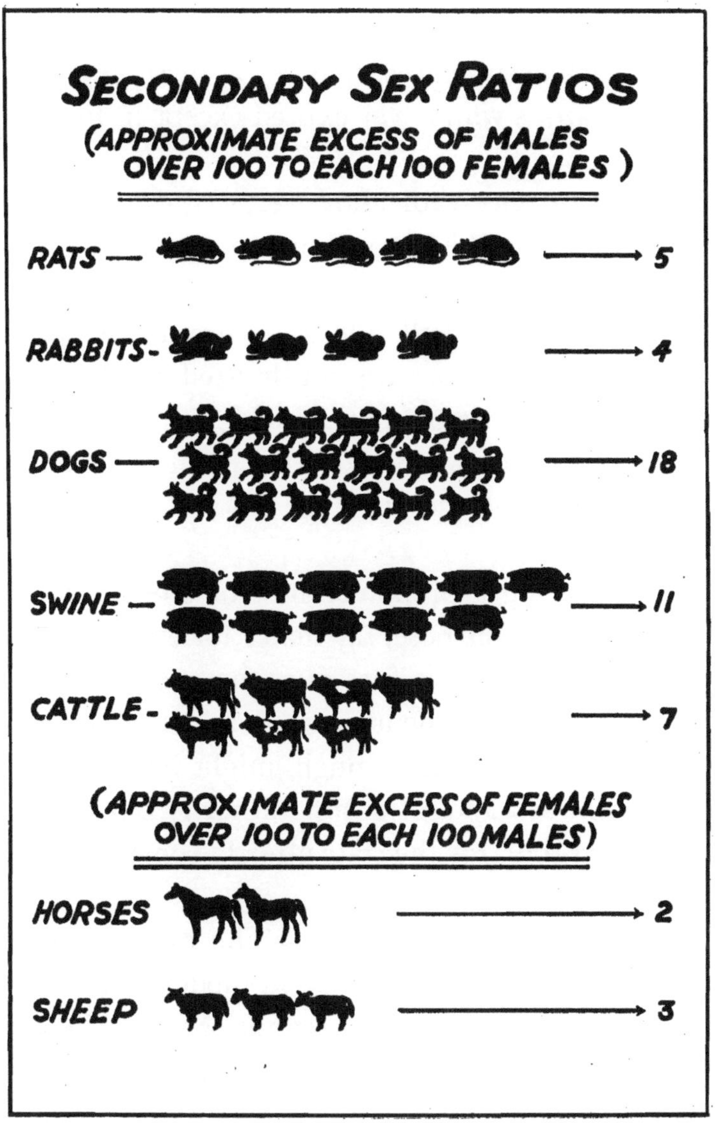

FIGURE 42.—AVERAGE MALE-FEMALE RATIOS AT BIRTH

There is an excess of males in rats, rabbits, dogs, swine and cattle, and an excess of females in horses and sheep.

poses. Such occurrences would be as mythological as the Amazons who never existed except in lively imaginations.

It is not inconceivable, however, that the public welfare might well be served by some governmental or group influence upon the sex of citizens-to-be. One example will suffice. Certain diseases, such as haemophelia, are transmitted solely by females. Would it be unreasonable for the state to try to keep the progeny of stricken mothers exclusively male and thereby wipe out the disease in one generation?

As for raising a preponderance of one sex for a specific-group purpose, say males for expanding a particular nation, it first must be proved that one form of government is good and the other bad. Sex control, true enough, might be used to some extent as a tool of group authority but only as an added tool, for the state already controls life and death.

The Parents Will Choose—The overwhelming verdict of both primitive and modern history is that mass manipulation of sex ratios would be a practical impossibility and that the ultimate and final choice would rest with the exclusive participants in the act of control—parents themselves.

Birth control and family limitation today are commonplace practices, but nowhere have governmental agencies succeeded by force in regulating the size of individual families. While governmental, religious and social groups may advise, the actual use of contraceptives or medical aids to fertility remain largely matters of personal choice. There is every reason to believe that the question of using sex control chemicals will likewise require a personal answer.

While it must be admitted that family limitation to some extent has come to stay, most potential parents desire children but are not necessarily satisfied with the arbitrary sex selections imposed upon them. It is a common experience to hear a mother exclaim: "I should like to have another baby if I only could be sure it would not be another girl (or boy)."

Assuming that a majority of parents could and will eventually choose the sex or sexes of their children, let us now see if we can understand the social forces and mental processes which would influence their decisions.

Wanted: Males to Rule and Inherit—Unquestionably in the past the demand for boys has been heard more frequently than the demand for girls. There is nothing mysterious about this apparent desire for males.

The male, because of his strength and aggressiveness, emerged as the leader and defender of the primitive family. Fighting with tooth and claw, club and stone, meant survival. Males were the best fighters. Parents, therefore, realized the need for a son to feed and protect the family group in the event of the father's disability or death.

The importance of at least one son, preferably the first-born child, was enhanced by the possessive instinct. Fathers wished to pass on their property to strong hands. Chiefs and later on monarchs needed male heirs to perpetuate dynasties. As the labors performed by women in primitive societies were assumed by men, boys were required to inherit a farm or carry on a business.

Among rare tribes the practical value of women was evidenced by the sale of girls in marriage. The Bantus, for example, regard wives as an economic asset and the bride-price or lobola is paid in cattle. Matriliny, the custom of reckoning succession and inheritance in the female line, survives in various parts of the world. Such special premiums on daughters, however, are exceptions to the rule that at least one son has been and still is desired in a majority of families.

Wanted: Females to Love and Protect—It will be noted that the man-child desire usually extends only to a single birth. One son in a family, one successor to a throne, one heir to a business, fulfills the requirement. Thereafter, it is a different story.

The demand is not for boys exclusively. The parents of one male child almost always want the second child to be a daughter and again the reasons, while less practical, are clear.

The mother wants a daughter for companionship, sympathy and understanding. The father instinctively desires a daughter, as he desired a mate, to protect and as an object of paternal affection. This is a desire born of sentiment but one fully as real and impelling as the pride which demands a son.

Practically all prospective parents desire children of both sexes. We all know parents who appear to have gone on having four, five or six undesired children in the hope of the arrival of a desired boy or girl. They say, "I want first a boy and then a girl," or "We have two boys now so we want a girl," or "We hope it's a boy. Three girls are too many." Mass interviews and polls uniformly sustain the axiom that parents with only boys want a change, and vice versa.

To perpetuate the human race each pair of adults must produce three children. The average family consists of four. It is logical to conclude that parental sex control would not materially alter the general proportions of the sexes, although it would adjust them—happily—within individual families.

The Balanced Family—What people want may not always be good for them, but certainly the case for all-girl or all-boy families is a poor one from the viewpoint of either parents or children.

Each sex tends to stimulate, liberalize and normalize the other. An all-girl family is apt to become femininized, an all-boy family masculinized. A mother surrounded by daughters misses the broadening influence of males, while a father surrounded by sons misses the softening influence of females.

Boys with sisters and girls with brothers come to take the opposite sex for granted and more quickly lose self-consciousness in the contacts of adolescence and maturity. The penalties paid by the sisterless boy or the brotherless girl can be avoided by planning a family with a fair number of children of each sex.

An answer to the question of what sex control will do to the human race can be given with at least a fair amount of assurance. It will provide a new and effective means of increasing the sum total of human happiness.

THE NEW SCIENCE ADVANCES

(AUTHOR'S NOTE: *Since this book came from the presses new developments have upheld the basic theory and broadened the scope of sex predetermination. Recent progress is outlined in the following supplementary pages. C. W.*)

Confirmation at Illinois—A history-making scientific verdict sustaining the claim that sex can be fixed before birth—one of the longest advances made toward giving millions of parents boys or girls as they choose—was delivered by a noted geneticist Nov. 30, 1940, before members of the American Society of Animal Production.

Dr. Elmer Roberts of the Department of Animal Husbandry, University of Illinois, reported a long series of successes in his laboratory as a result of using the acid-alkali method in experiments with more than 1,800 rats.

His paper concluded decisively: "Douching females with lactic acid before mating was effective in increasing the number of females, while the use of sodium bicarbonate increased the number of males."

The Roberts report came as strong confirmation of the work done at the Applied Research Laboratories, Dayton, N. J., where biologists, sponsored by *The New York Daily News,* conducted the first large scale sex control tests in the United States.

Dr. Roberts not only obtained heavy preponderances of males with soda and of females with acid, but also reversed the treatments successfully 23 times. His final figures were:

(Acid Group)

Pct. Sol.	No. of Litters	M.	F.	Sex Ratio (Males to 100 Females)
3%	69	166	301	55.1
2%	32	111	163	68.1
Totals	101	277	464	59.7

(Soda Group)

Pct. Sol.	No. of Litters	M.	F.	Sex Ratio (Males to 100 Females)
3%	35	157	126	124.6
5%	70	366	177	206.8
Totals	105	523	303	172.6

During one phase of his experiment Dr. Roberts recorded 61 straight litters, all with majorities of the sex indicated by the treatment. Such a succession, he calculated, could occur naturally only "once in a million billion times."

The Roberts Technique—The University of Illinois scientists described his method as "essentially similar" to that used by Thomas Durfee at the Applied Research Laboratories as set forth in this book.

The treatment with most litters consisted of douching the vagina 3 times in immediate succession by means of a medicine dropper. All matings occurred within a period of 2 hours from the time of douching. Dr. Roberts said, however, that he was unable to aspirate the solutions in and out as suggested by Durfee.

Like Durfee, he found that 3 per cent solutions of acid were more effective than 2 per cent and 5 per cent solutions of soda more effective than 3 per cent. He reported that the increased strengths "may have slightly reduced litter size," but "of

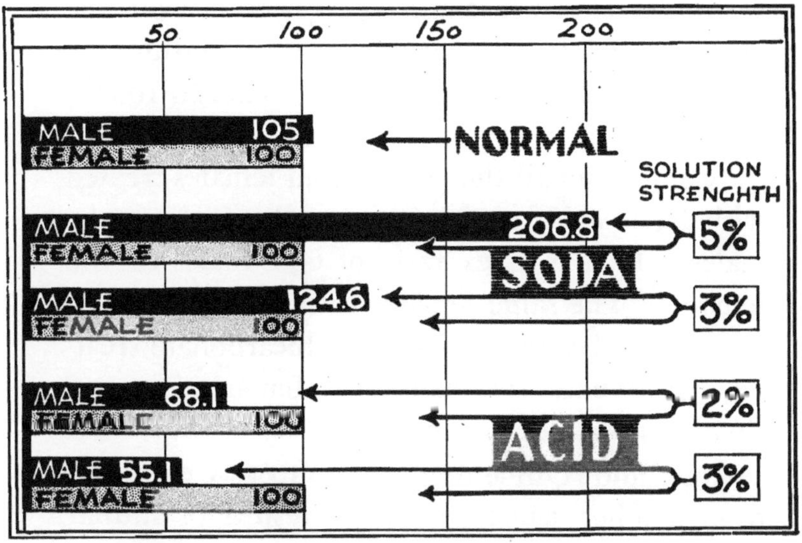

FIGURE 15.—UNIVERSITY OF ILLINOIS RESULTS
The chart shows how Dr. Roberts influenced the sex of experimental rats with soda and acid solutions.

this we cannot be sure because the animals were of different ages when the litters were produced."

For his "reversals" Dr. Roberts selected several mother rats which previously had been treated successfully with acid and switched the treatments with the following results:

"With 2 per cent lactic acid 11 litters with 98 young showed a sex ratio of 84.9 males to 100 females. The same females with 3 per cent sodium bicarbonate had a sex ratio among 98 young of 127.9 males to 100 females. With 3 per cent lactic acid 171 young showed a sex ratio of 61.3 males to 100 females, while among 163 young from the same females treated with 5 per cent sodium bicarbonate the sex ratio was 207.5 males to 100 females.

"Combining all the young from females treated with lactic acid, we have 110 males and 159 females, which is a sex ratio of 69.2 males to 100 females. The same females with the combined 3 per cent and 5 per cent sodium bicarbonate treatments produced 165 males to 96 females, a sex ratio of 171.9 males to 100 females."

Cole and Quisenberry Results—Geneticists from 2 other schools reported results of similar tests to the Society. The papers were presented by Dr. Leon J. Cole of the University of Wisconsin and Dr. John H. Quisenberry of the Agriculture and Mechanical College of Texas.

At Wisconsin sex ratios were not materially altered and the Cole verdict was "not proven." At Texas A. & M. the solutions failed to influence rat ratios, but the treatments were followed by unusual ratios in rabbits.

Dr. Cole reported on douching treatments followed by a total of 271 litters of rats and 35 litters of rabbits. The sex ratios ranged only from 47.6 to 54.1 which were not significant changes from normal.

Dr. Quisenberry obtained a total of 110 litters of rats without significant deviations from normal. Then he turned to rabbits and the picture changed.

He obtained 8 litters from female rabbits douched with 3 per cent soda. There were 21 males to 26 females. Using a 5 per cent solution he got 32 males to 17 females in 8 additional litters. With 2 per cent acid he obtained 16 litters with 32 males and 56 females.

The 3 scientists, after presenting their papers, agreed that further studies should be made in an effort to find an explanation of the different results in the strains of animals, variations in technique, or some other condition.

Human Successes in America—It was inevitable that physicians in the United States would follow the lead of Unterberger in Germany and the Taylors in England and attempt application of the acid-alkali method to humans. They have done so with impressive results.

Prior to this writing 5 American doctors reported successes in obtaining babies of the sex desired. They were Dr. Mor Lewinski, 309 W. 99th St., New York, N. Y.; Dr. Nicholas E. Caputo, 120 Franklin Ave., Brooklyn, N. Y.; Dr. J. M. Behar, 1266 Grand Concourse, Bronx, N. Y.; Dr. M. M. Marcus, 1318 42d St., Brooklyn, N. Y.; and Dr. Sidney B. Callis, 897 S. Main St., Warren, R. I.

Also reported was the successful use of both acid and soda solutions by a family which followed instructions mailed by Dr. Harold Taylor from London.

To obtain a girl they purchased from a pharmacy 27.8 c.c. of 85 per cent lactic acid. This was dissolved in 2 pints of boiled lukewarm water. The fluid was used as a douche 2 hours before attempted conception. To obtain a boy they bought 360 grains of pure reagent grade sodium bicarbonate. This was mixed with water in the same way as the acid, and a son resulted.

Dr. Taylor previously had successfully advised at least 3 other American couples how to use the technique.

The activity of the English pioneer, however, was curtailed by the war and the virtual blackout of non-military scientific research overseas. Writing to *The News,* Dr. Taylor said he considered the Dayton tests "very convincing." He added:

"You must try and persuade your energetic young men to carry on the work where we have left off. But let's hope that the boys they produce will be luckier than those of our generation."

Dr. Taylor and his brother, Dr. Gordon Kirwan-Taylor, who worked together in England to achieve 84 per cent success in producing sons or daughters as desired by the mothers, both volunteered for service in the British Army.

Monkeys and Humans—Discovery of the fact that the acid-alkali method will work on monkeys was reported by Dr. Callis, who also found it a complete success in 10 human cases.

The Rhode Island doctor revealed that he obtained male and female offspring at will from Macacus rhesus or Bengal monkeys. Later he advised 10 women patients. Six wanted and had girl babies. Four wanted and had boys. There were no failures.

Dr. Callis was the first scientist in America—probably in the world—to use the technique on the highest order of mammals next to man. While his remarkable work did not come to light until

1941, he had carried it on for several years, starting in 1935. Like Unterberger he first noticed that soda douches used to overcome sterility usually resulted in male offspring. He deduced that acidity should produce the opposite sex.

First he experimented with rats. Then he controlled the sex of a dog litter. His success led him to buy 4 pairs of monkeys, animals most closely resembling humans in that the females have a 28-

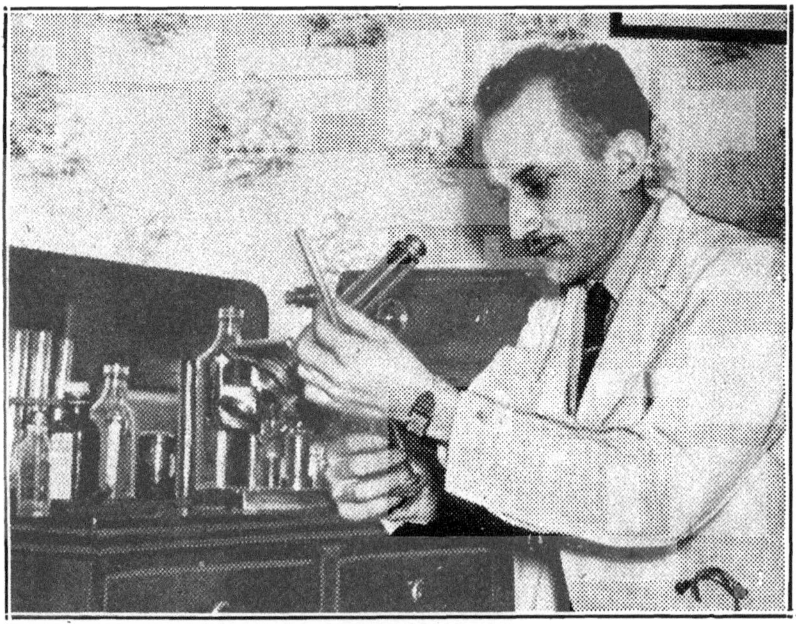

FIGURE 28.—SUCCEEDED WITH MONKEYS
Dr. Sidney B. Callis, of Warren, R. I., who applied the acid-alkali method to the highest order of mammals next to man.

day menstrual cycle and a pregnancy period of 9 months.

A female monkey, "E-Z-M," bred normally, bore a male. Another female treated with a 2 per cent acid solution bore a female. "E-Z-M," acid-treated, then bore a female and, soda-treated at her next mating, bore a male.

Meanwhile, Dr. Callis began a study of the pH of the human vaginal tract and of semen. He found that vaginal secretions normally are acid, ranging from 3.86 to 4.45. He found the pH to be altered by the cervical secretion, usually acid. He set the normal pH of the spermatic fluid at 7.6. At any pH below 2.0 the sperms apparently were paralyzed.

With this groundwork laid, he ascertained that a pH of 8.0 in the vaginal tract is best for boy conception and of 5.5 to 6.0 for girl conception. This coincided almost exactly with the Taylors' figures —8.0 and 5.5—arrived at independently.

Technique of Dr. Callis—Finally, Dr. Callis tackled the job of controlling the pH in women patients and thereby fixing the sex of their babies. For a girl he advised the patient to take the following steps:

(1) Ninety minutes before attempted conception douche with a solution made by dissolving

27.8 c.c. of 85 per cent lactic acid in 2 pints of boiled lukewarm water.

(2) Soak the solution in a 2-inch tampon or rubber sponge with a string attached. Insert immediately after first douche and remove in ½ hour.

(3) Douche a second time 1 hour before attempted conception.

To obtain a boy he advised the same procedure except that the solution consists of 360 grains of pure reagent grade sodium bicarbonate dissolved in 2 pints of water.

Essentially the Callis method was the same as that recommended by Dr. Taylor and other physicians. However, instead of a single douche, Dr. Callis advised the tampon and second douche to insure the right pH. He said he considered the time element important and insisted that his patients set an alarm clock and time the douches "to the minute."

Fifteen Controlled Births—A phenomenal demonstration of predetermination—the first large scale achievement with humans in America—was attained by Dr. Caputo. He scored 87 per cent success based on 15 cases.

Thirteen women patients under his observation obtained the sex of child they desired. They gave birth to 8 sons and 5 daughters. In 2 cases there were failures, apparently due to carelessness in using the treatment recommended.

The boy-bearing mothers used a 2 per cent solution of sodium bicarbonate while the mothers of girls used a 3 per cent solution of lactic acid. The water used for mixing the solutions was ordinary warm water. Several of Dr. Caputo's patients used the solutions as douches from 1 to 4 months prior to conception.

"I am convinced that the method will succeed in 75 to 80 per cent of the cases," said Dr. Caputo. "I believe it works unless a mistake is made in technique."

Progress With Cattle—Animal geneticists and stockmen all over the country watched the results of tests of the acid-alkali method with dairy cattle at the Shelby County Penal Farm, Memphis, Tenn. (See Page 148.) With 26 calves born, 16 were of the sex determined by the breeders before mating. Wallace Bryan, supervisor, reported the results as follows:

(Acid Group)

Number of cows treated	20
Males born	7
Females born	13

(Alkali Group)

Number of cows treated	6
Males born	3
Females born	3

While no apparent success was achieved in the alkali group, results in the larger acid group seemed to be either decisive proof of the theory or a near-miracle. The normal sex ratio in cattle is 107.3 males to 100 females. Calves were born to the acid-treated cows in ratio of 50 males to 100 females. Odds against such a violent upset occurring by chance are approximately 169 to 1.

All of the calves in the test groups were healthy, normal and in every way as valuable as other animals produced in the herd. The first sex-controlled bull calf sold for $500 before birth.

The Record in Detail—Bryan kept careful records to show the solution strengths and time elapsing between treatment and insemination. Calves arrived in the following order:

(Acid Group)

Name of Cow	Pct. Sol.	Minutes Elapsed	Sex of Calf
Volunteer Bellflower	1.0	60	Bull
Volunteer Cherry Switch	.5	60	Bull
Afterglow Pet	.5	60	Heifer
Signal Lula Bell	.5	140	Bull
Afterglow Sis Ora	1.0	180	Heifer
Signal Favorite Lady	1.0	120	Heifer
Volunteer Ednaerl	1.0	120	Heifer
Volunteer Glamorous Blonde	1.0	120	Heifer
Signal Volunteer Carolyn	1.0	120	Bull
Volunteer Signal Gem	1.0	124	Heifer
Afterglow Little Maid	1.0	120	Bull

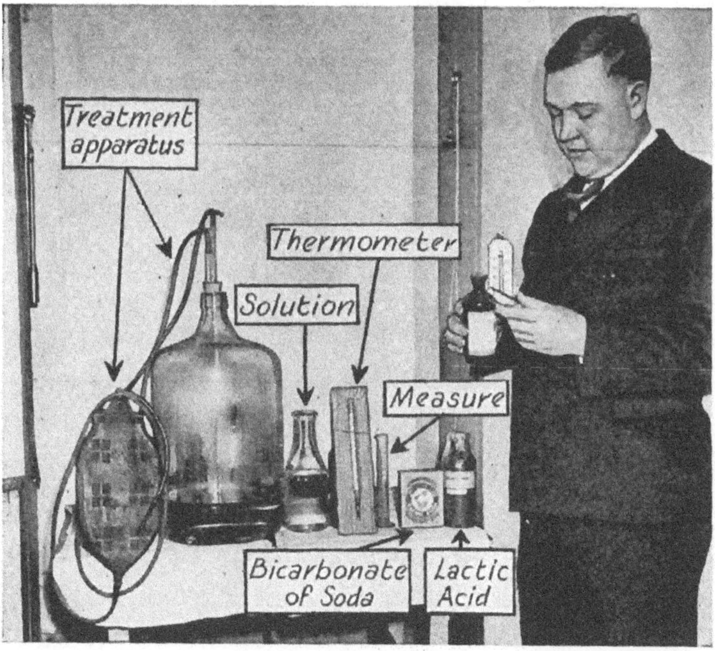

FIGURE 38.—APPARATUS FOR CATTLE

Wallace Bryan, supervisor of cattle sex control tests at Shelby County Penal Farm, with chemicals and equipment used in experiment.

Name of Cow	Pct. Sol.	Minutes Elapsed	Sex of Calf
Volunteer Winnie Katherine ...	1.0	120	Heifer
Observer Welcome Lady	1.0	95	Heifer
Observer Precious Gem........	1.0	135	Heifer
Signal Wallflower	1.0	165	Heifer
Bet Observer Lassie	1.0	60	Bull
Signal Knight Oxford Lady	1.0	120	Heifer
Afterglow Dreamy Kiss	1.0	120	Heifer
Afterglow Miss Bruce	1.0	15	Heifer
Volunteer Ednaerl	1.0	10	Bull

(Alkali Group)

	Pct. Sol.	Minutes Elapsed	Sex of Calf
Volunteer Straight Lines	2.0	40	Bull
Doubles Wallflower	2.0	55	Bull
Volunteer Bravo Bouquet	2.0	120	Heifer
Signal Sybil Queen............	2.0	115	Heifer
Signal Pretty Poppy Pansy	2.0	200	Bull
Volunteer Sybil Lass	2.0	105	Heifer

Analysis of the Memphis records indicated that the proper solutions and the right amount of time elapsed between douching and conception may be important factors affecting the percentage of success. Solutions too strong seem likely to prevent conception. Those too weak are ineffective. The vaginal pH of cows at estrus is near 7.0 or neutral. A pH of about 8.0 probably is favorable for bull conception and of about 5.0 or 6.0 for heifer conception, but no study has been made as yet to set these figures with precision.

More Cattle, Horses and Dogs—For the first time in the history of the cattle country during 1940 a dairyman, Robert D. Lindley, of Yantis, Tex., was able to predetermine the sex of calves 9 times in succession without a failure.

Discouraged by the number of bull calves he was getting and hearing of the Memphis work, Lindley treated 48 cows in his Jersey herd with a

1 per cent lactic acid solution. Each of the first 9 to bear calves dropped heifers.

Lindley used lukewarm rain water to mix his solution. He douched each cow with a syringe tube 60 minutes before natural service. The experiment cost him only $5.

On the heels of the report of Lindley's remarkable results, authentic new evidence that horses respond to the technique came from Kentucky. Joe H. Palmer, associate editor of the *Blood-Horse,* published by the American Thoroughbred Breeders' Association at Lexington, Ky., reported the results of a test conducted by a physician-breeder.

The breeder found that 4 out of 9 of his mares failed to be in foal in the 1939 season. Tests with litmus paper showed the vaginal tracts of the 4 to be acid. He rinsed them with soda, bred them again and this time each got in foal. Three of the mares gave birth to colts. Only 1 bore a filly.

While these new experiments with large animals progressed steadily but slowly, sex control in dog breeding became comparatively common place. So many successes were reported during 1940 and 1941 that space here does not permit their discussion. (See page 134.)

First Swine Success in America—More economic production of pork was foreshadowed in

1941 by the first success in this country in adapting the acid-alkali method to swine. It was achieved by Lloyd A. Truesdale, of Lafayette, O., owner of a prize-winning herd of Hampshire hogs.

For the test Truesdale used Gay Juliet 1st, a daughter of Juliet, grand champion at the Ohio State Fair of 1937-38, and Gay Score, one of the great sires of the breed.

"I treated with a 3 per cent soda solution," Truesdale said. "I used 2 quarts of boiled rain water in an ordinary hot water bottle."

The sow gave birth to a litter of 9 male and 3 female offspring—a 3 to 1 male ratio. Truesdale thereafter treated half of his brood sow herd with 3 per cent acid douches with a view to obtaining females.

The swine was the eighth species of mammal to which the technique was known to have been applied with positive results since experiments sponsored by *The New York Daily News* were started in 1938. Other mammals to which the method had been successfully adapted were human beings, monkeys, horses, cattle, dogs, rabbits and rats.

GLOSSARY

ASEXUAL—Having no sex.

CELL—A minute mass of protoplasm usually having a nucleus and inclosed in a covering or cell-wall.

CERVIX—The neck of the womb.

CHROMOSOME—Substance within living cells which determines characteristics of the living matter, therefore the factor which determines sex.

CONCEPTION—The act of becoming pregnant or state of being conceived.

CYTOPLASM—The protoplasm of a cell exclusive of the nucleus.

EGG—The ovum.

ESTRUS—The period of heat or season when the female will accept the male.

FERTILIZATION—Impregnation or fecundation.

FOLLICLE—A small cavity or gland.

GAMETE—A male or female reproductive cell.

GESTATION—The period of pregnancy.

LITTER—A number of young brought forth at one time by a multiparous animal.

MAMMAL—One of the highest class of vertebrates, including man and all other animals that suckle their young.

MENSTRUATION—The discharge of the menses.

NUCLEUS—A mass of protoplasm in the interior of cells forming an essential element in growth and reproduction.

OVARY—The female reproductive gland in which the eggs are produced.

OVULATION—The formation in and departure of eggs from the ovary.

OVUM—The egg or female reproductive cell.

PROTOPLASM—The essential substance of the cell body.

SEMEN—The fluid containing sperms produced by the male.

SPERM—The reproductive cell produced by the male.

STERILITY—The inability to produce living young.

TESTICLE—The male organ which produces sperms.

UTERUS—The womb of the female.

VAGINA—The lower end of the reproductive canal of the female into which the uterus opens.

WOMB—The uterus.

ZYGOTE—The cell formed by the union of two gametes.

INDEX

Milton Keynes UK
Ingram Content Group UK Ltd.
UKHW010013020324
438794UK00002B/23